T0192609

Molecular Modeling in Heavy Hydrocarbon Conversions

CHEMICAL INDUSTRIES

A Series of Reference Books and Textbooks

Consulting Editor

HEINZ HEINEMANN
Berkeley, California

Molecular Modeling in Heavy Hydrocarbon Conversions

Michael T. Klein
Gang Hou
Ralph J. Bertolacini
Linda J. Broadbelt
Ankush Kumar

CRC Press
Taylor & Francis Group
Boca Raton London New York

CRC Press is an imprint of the
Taylor & Francis Group, an **informa** business
A TAYLOR & FRANCIS BOOK

Published in 2006 by
CRC Press
Taylor & Francis Group
6000 Broken Sound Parkway NW, Suite 300
Boca Raton, FL 33487-2742

First issued in paperback 2020

ISBN 13: 978-0-367-57802-2 (pbk)
ISBN 13: 978-0-8247-5851-6 (hbk)

Library of Congress Card Number 2005048510

Library of Congress Cataloging-in-Publication Data

Molecular modeling in heavy hydrocarbon conversions / by Michael T. Klein ... [et al.].
    p. cm. – (Chemical industries series ; 109)
    Includes bibliographical references and index.
    ISBN 0-8247-5851-X
    1.Hydrocarbons – Mathematical models. 2. Molecular Structure – Mathematical models. 3. Chemical Kinetics – Data processing. I. Klein, Michael T. II. Chemical industries ; v. 109.

QD305.H5M54 2005
547'.01'0151—dc22   2005048510

Visit the Taylor & Francis Web site at
http://www.taylorandfrancis.com

and the CRC Press Web site at
http://www.crcpress.com

Dedications

To Dorothy for her encouragement and patience
Ralph J. Bertolacini

To Jim, Jenna, and Daniel
Linda Broadbelt

To Ping and Alex
Gang Hou

To Bitsy, Jenny, Michael, and Lisa, with love
Michael T. Klein

Preface

Molecular Modeling in Heavy Hydrocarbon Conversions is the result of the contributions of many colleagues. I'd like to use this Preface to recognize and thank them all.

The research program that links these colleagues began at the University of Delaware in 1981 and continued at Rutgers University in 1998. Its principal philosophy developed in P. S. Virk's lab at MIT during the 1970s and 1980s, this research program began as a blend of experimental work, aimed at discerning the reaction pathways underlying the reactions of complex systems, and modeling work, aimed at packaging the experimental insights into a quantitative summary. The program flourished and, by 1990, many complex systems had come under investigation. At this time, we began to realize that, in our modeling work, we were, essentially, repeating ourselves every time we developed a new kinetic model. This led us to attempt to formalize the modeling approach, and, ultimately, to capture this approach in the form of a computer program that built other computer programs, i.e., model building software. The generic features of this model building capability are described in Chapters 1 to 6 and the remaining chapters are devoted to a handful of reasonably comprehensive applications.

Molecular Modeling in Heavy Hydrocarbon Conversions is, in this sense, the combined product of our colleagues Martin Abraham, Brian Baynes, Craig Bennett, Nazeer Bhore, Ken Bischoff, Lori Boock, Jim Burrington, Darin Campbell, Michel Daage, Stavroula Drossatou, Dean Fake, Bill Green, David Grittman, Cindy Harrell, Frederic Huguenin, Sada Iyer, Bill Izzo, Steve Jaffe, Prasanna Joshi, Michael T. Klein, Jr., Stella Korre, Concetta LaMarca, Ralph Landau, Tom Lapinas, Mike Lemanski, Cristian Libanati, Dimitris Liguras, Tahmid Mizan, Sameer Nandiloya, Matt Neurock, Abhash Nigam, Giuseppe Palmese, Frank Petrocelli, Tom Petti, Bill Provine, Richard Quann, Carole Read, Don Rohr, Carlonda Russell, Stan Sandler, Shalin Shah, John Shinn, Scott Stark, Ryuzo Tanaka, Susan Townsend, Pete Train, Dan Trauth, Achin Vasudeva, Preetinder Virk, Tim Walter, Xiaogong Wang, Beth Watson, Bob Weber, Wei Wei, Ben Wu, and Musaffer Yasar. The five co-authors who assembled the manuscript would like to acknowledge them all as contributing scholars, and recognize that the final manuscript is a cumulative product that has an intellectual element of them all in it.*

---

* I would like to acknowledge the specific contributions of former students for whom I served as a research advisor and whose papers and thesis chapters have provided substantial material for this book: Darin Campbell, Dan Trauth, and Tom Petti for Chapter 2; Prasanna Joshi for Chapters 3, 7, and 11, Stella Korre and Matt Neurock for Chapter 4.

I would also like to acknowledge the superb intellectual environments at the University of Delaware and Rutgers, The State University of New Jersey, that allowed this work to develop and be assembled.

Michael T. Klein

Authors

Ralph J. Bertolacini is currently an independent consultant and special term appointee at Argonne National Laboratory. After 39 years with Amoco, with experience in analytical, inorganic, and catalytic chemistry, he retired as director of exploratory and catalysis research. He was a charter member of the Center for Catalysis Science and Technology, and adjunct professor of Chemical Engineering at the University of Delaware. The author of over 25 technical papers and 86 US patents dealing with petroleum refining, he received his B.S. degree from the University of Rhode Island and an M.S. degree in chemistry from Michigan State University. He was named a Michigan State Distinguished Alumni in 1991. Mr. Bertolacini was charter member and past president of ASTM Committee D-32-Catalysis, and in 1987, the recipient of the Eugene Houdry Award in Applied Catalysis presented by the North American Catalysis Society. He is active in the Chicago Catalysis Club, and in 1993 was awarded the Ernest Thiele Award presented by the Chicago section of the American Institute of Chemical Engineers (AICHE).

Linda Broadbelt is a professor in the Department of Chemical and Biological Engineering at Northwestern University. She received her B.S. degree in chemical engineering from The Ohio State University and graduated *summa cum laude*. She completed her Ph.D. degree in chemical engineering at the University of Delaware where she was a Du Pont Teaching Fellow in Engineering. At Northwestern, she was appointed the Donald and June Brewer Junior Professor from 1994–1996. Professor Broadbelt's research and teaching interests are in the areas of multiscale modeling, complex kinetics modeling, environmental catalysis, novel biochemical pathways, and polymerization/depolymerization kinetics. A major emphasis of her research is the computer generation of complex reaction mechanisms, and application areas include biochemical pathways, silicon nanoparticle production, and tropospheric ozone formation. Professor Broadbelt is associate editor for *Energy and Fuels* and currently serves as the chair of programming for the Division of Catalysis and Reaction Engineering of AIChE. She was appointed to the Scientific Organizing Committee for the 19th International Symposium on Chemical Reaction Engineering and has served on the Science Advisory Committee of the Gulf Coast Hazardous Substance Research Center since 1998. Dr. Broadbelt's honors include a CAREER Award from the National Science Foundation, appointment to the Defense Science Study Group of the Institute for Defense Analyses, and selection as the Ernest W. Thiele Lecturer at the University of Notre Dame and the Allan P. Colburn Lecturer at the University of Delaware.

Gang Hou is a senior director of consulting at Unica Corporation, a leading enterprise marketing management software firm. Prior to this position, he was a visiting professor of engineering at Rutgers, The State University of New Jersey. Before becoming a visiting professor, Gang Hou was the lead solution strategist responsible for the e-marketplace operations at i2 Technologies, a leading supply chain management software firm. Dr. Hou received a B.S. degree with a double major in polymer science and applied mathematics from East China University of Science and Technology, and an M.S. degree in computer science and Ph.D. degree in chemical engineering from the University of Delaware. He is working on his M.B.A. degree in entrepreneurship at Babson College. Dr. Hou has consulted for many blue-chip firms, including Accenture, Corporate Express, Discover, E*Trade, IBM, JP Morgan Chase, and MBNA, regarding their business strategy and technology implementation. He conducts research in the interface between chemical engineering and computer science, with a special interest in the kinetic modeling of complex systems.

Michael T. Klein is the Dean and Board of Governors Professor of Engineering at Rutgers, The State University of New Jersey. Previously, Professor Klein was the Elizabeth Inez Kelley Professor of Chemical Engineering at the University of Delaware, where he also served as Department Chair, Director of the Center for Catalytic Science and Technology, and Associate Dean. Professor Klein received his BChE degree from the University of Delaware in 1977 and his Sc.D. degree from MIT in 1981, both in chemical engineering. The author of over 200 technical papers, he is active in research in the area of chemical reaction engineering, with special emphasis on the kinetics of complex systems. He is the Editor of the ACS journal *Energy and Fuels* and the Reaction Engineering Topical Editor for the *Encyclopedia of Catalysis*. He serves on the Editorial Board for *Reviews in Process Chemistry and Engineering* and the McGraw-Hill Chemical Engineering series. Dr. Klein is the recipient of the NSF PYI Award and the ACS Delaware Valley Section Award.

Table of Contents

Part I
Methods

Part II
Applications

1 Introduction

1.1 MOTIVATION

It is difficult to model the real world because of its complexity. The enormous complexity of various chemical reaction systems that this work focuses on has historically defied fundamental analysis. This book introduces, develops, integrates, and formalizes a detailed systematic molecule-based kinetic modeling approach and a system of chemical engineering software tools to delineate and reduce the essential elements of the complexity in the modeling of complex reaction systems. This approach is then applied to the development of molecular models in heavy hydrocarbon conversion processes including catalytic reforming, hydrocracking, hydrotreating, hydroprocessing, and fluid catalytic cracking (FCC), as well as thermal cracking (pyrolysis) in the petroleum refining industry.

Because of limitations mostly in the analytical chemistry and computer hardware and software capabilities, most traditional and current process models have implemented lumped kinetics schemes, where the molecules are grouped by global properties such as boiling point or solubility. About the order of 10 lumps have generally been used to represent the complex feedstock and reaction systems. Even in recent years, more complex kinetic models are scarcely used in the industry (Bos et al., 1997). Molecular information is thus obscured because of the multicomponent nature of each lump. This approach unavoidably leads to the absence of properties that are beyond the definition of lump because of the absence of chemical structure. The thus developed globally lumped and nearly "chemistry-free" kinetic models are specific in nature and cannot be extended to the new feedstocks and catalysts.

However, both increasing technical (such as product performance) and environmental (such as the Clean Air Act) concerns have focused attention on the molecular composition of petroleum feedstocks and their refined products. For example, recent environmental legislation has placed restrictions on the maximum allowable benzene content in gasoline and sulfur content in diesel. Thus, the new paradigm is to track each molecule in both the feed and product throughout the process stream.

Molecules are the common foundation for feedstock composition, property calculation, process chemistry, and reaction kinetics and thermodynamics. Molecule-based models can incorporate multilevel information from the surface and quantum chemical calculations to the process issues and can serve a common fundamental form for both process and chemistry research and development. Modeling approaches that allow for reaction of complex feeds and prediction of molecular properties require an unprecedented level of molecular detail.

Two enabling technological advancements have helped modeling at the molecular level become achievable. First, recent developments in analytical chemistry now permit the direct, or at least indirect, measurement of the molecular structures in complex feedstocks. Second, the advancement in information technology, especially the explosion of computational power, allows for the necessary documentation to track the fate of all the molecules during both reaction and separation processes. Collectively, both the strategic forces on rigorous models and the enabling analytical and computational advances motivate the development of molecule-based detailed kinetic models of complex processes.

The construction of detailed kinetic models is complicated by the large number of species, reactions, and associated rate constants involved. Modern analytical measurements indicate the existence of $O(10^5)$ unique molecules in petroleum feedstocks. Each species corresponds to one equation in a rigorous deterministic approach; therefore not only the solution but also the building of the implied model is formidable. Keeping track of $O(10^5) \times O(10)$ reactions manually is impractical and too complicated to do in a time- and cost-efficient fashion. This has motivated the development of a system of software tools to automate the entire model building, solution, and optimization process, thereby allowing process chemists and engineers to focus on the process chemistry and reaction kinetics by using the software tools to do the human error-prone and repetitive work accurately and quickly.

1.2 BACKGROUND

Table 1.1 summarizes the kinetic modeling approaches at various levels. In the lumped approach, all the feedstocks, reaction networks, and the products are defined as global lumps and hence lack fundamental kinetic information and predictive capability. Developing a model with detailed kinetic information requires modeling of the chemistry at either the pathways level or the mechanistic level.

At the pathways level, the model contains most of the observed species explicitly and describes the molecule-to-molecule transitions in the reaction network. The reaction mechanism implicitly guides the model development in terms of both the reaction network and rate laws. The formulation of rate laws at the pathways level involves many *a priori* assumptions, such as the rate-determining step (RDS). The corresponding mathematical model is numerically friendly and can be solved quickly compared with the corresponding mechanistic model.

TABLE 1.1
Different Levels of Kinetic Modeling

Model Type	Model Description	Characteristics
Lumped	Measurable lumped groups	Feedstock dependent Lacks predictive capability
Detailed at pathways level	Observable molecules	Feedstock independent Approximate rate constants
Detailed at mechanistic level	Intermediates and molecules	Feedstock independent Fundamental rate constants

At the mechanistic level, the model contains a detailed explicit description of the mechanism, including both the molecules and intermediate species, such as the ions and radicals. Fewer *a priori* assumptions (such as RDS) are needed, so the rate parameters are more fundamental in nature. However, the corresponding mathematical model is more difficult to solve because of its inherent mathematical stiffness. Both the molecule-based detailed kinetic modeling approaches have the promise of obtaining feedstock-independent models and can be extrapolated to different catalysts in the same family.

Figure 1.1 shows the complexity of detailed kinetic models and a quantitative comparison between the pathways-level and mechanistic models with respect to

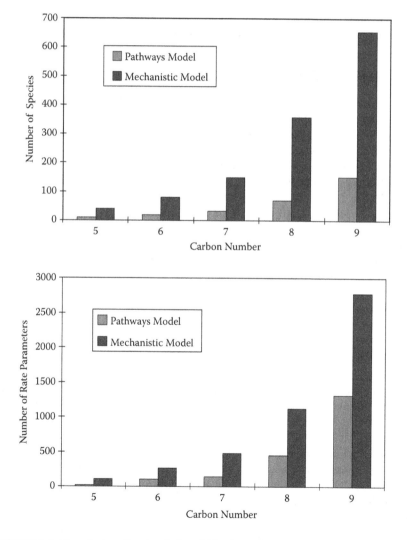

FIGURE 1.1 Complexity of molecule-based kinetic modeling.

the number of species and the number of reactions or rate parameters to be estimated for the complex reaction systems at the naphtha range. The number of species and the number of rate parameters increase exponentially with respect to the carbon number in both the pathways and mechanistic models, with the latter being far more sensitive. Simple feeds at the naphtha range can give complex models at the mechanistic level, and more complex feeds can give complex models at both the pathways and mechanistic levels. However, the complexity of the kinetic models that we are able to handle is balanced by the availability of the data, the limitation of the analytical chemistry, the computational power in terms of the CPU and memory, the mathematical methods for model solution and optimization, as well as the needs. A practical model would thus be an optimal subset of all feasible reactions that captures the essential chemistry of the process, although the complexity may increase as the modeling resources increase.

1.3 MODELING APPROACHES

The modeling of complex process chemistries such as thermal cracking, catalytic reforming, catalytic cracking, hydrocracking, hydrotreating, hydroprocessing, and FCC has taken decades to evolve. The initial models and modeling approaches were dictated by the limitations of the analytical characterizations. The early modeling approaches for thermal cracking (van Damme et al., 1975; Sundaram and Froment, 1977a,b, 1978;), catalytic reforming (Kmak et al., 1971; Ramage et al., 1987; Marin and Froment 1982; Mudt et al., 1995), catalytic cracking (Weekman et al., 1969; John and Wojciechowski, 1975; Jacob et al., 1976), hydrocracking, hydrotreating, and hydroprocessing (Qader and Hill, 1969; Stangeland, 1974; Laxminarasimhan et al., 1996) present various lumping strategies (Weekman, 1979; Astarita and Ocone, 1988; Aris, 1989; Gray, 1990). The lumped kinetic modeling approach often suffered from many drawbacks, and the lumped models were specific in nature and could not be extrapolated to different feedstocks and process configurations. These models often lacked mechanistic insights and hence could not be used to interpret the effects of catalyst properties and operating conditions. Finally, the changes in the composition of lumps in terms of molecular components often masked the true kinetics.

In the past two decades, more modeling efforts have gradually incorporated more molecular and structural detail in response to environmental and technical concerns. The fundamental hydrocarbon pyrolysis modeling conducted at the mechanistic level by Dente et al. (1979) and the carbon center modeling for catalytic cracking conducted at the pathways level by Liguras and Allen (1989a,b) are classic examples of detailed kinetic modeling for complex process chemistries. However, these elegant modeling approaches are not automated, and hence, it is tedious to rebuild and model complex processes containing thousands of species and reactions.

In the area of automated detailed kinetic modeling for complex process chemistries, the most comprehensive and elegant work includes the structure-oriented lumping (SOL) approach developed at Mobil (Quann and Jaffe, 1992, 1996) and the single-event approach developed by Froment and coworkers (e.g., Baltanas and Froment, 1985; Clymans and Froment, 1984; Hillewaert et al., 1988). The SOL approach uses vectors for the structural groups of molecules whose atoms are not explicit. The single-event approach is graph-theoretic oriented and can build fundamental kinetic models at the mechanistic level. The computationally intensive model building uses Boolean matrix multiplication (A^2 for one transition, A^3 for two transitions, etc.) to carry out chemical reactions.

Broadbelt et al. (1994) developed an automated computer-generated modeling approach for simple model compound (ethane) pyrolysis at the mechanistic level. This approach utilizes graph theoretic concepts for generation of the reaction network at both the pathways and mechanistic levels by representing molecules as atomically explicit bond–electron matrices and reactions as matrix operations. This approach uses matrix addition operations to carry out chemical reactions, which are much less CPU intensive and memory demanding. This approach is thus fast enough to allow the modeler to compare various pathways and mechanisms, insights, approximations, and their sensitivities to the final result within a short amount of time. The long-term goal of this modeling program was to integrate the various chemical engineering tools for the building, solution, and delivery of detailed kinetic models into one user-friendly software package.

1.4 MOLECULE-BASED KINETIC MODELING STRATEGY

Figure 1.2 summarizes the molecule-based kinetic modeling strategy used in this work. In the real world, the modeling goal is to predict product properties or the required operating conditions for a target set of product properties from the feed characterization. The kinetic modeling strategy provides an alternative route to achieve this goal at the molecular level, since molecules are the common foundation for feedstock composition, property calculation, process chemistry, and reaction kinetics and thermodynamics.

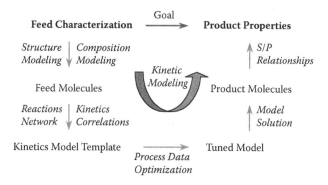

FIGURE 1.2 The molecule-based kinetic modeling strategy.

This approach begins with the molecular structure and composition modeling that uses stochastic simulation techniques to assemble a molecular representation of complex feedstocks from analytical chemistry, for example, H-to-C ratio, SIMDIS (Simulated Distillation), NMR(Nuclear Magnetic Resonance). Then, graph theory techniques are utilized to generate the reaction network. Reaction family concepts and quantitative structure reactivity correlations (QSRCs) are used to organize and estimate kinetic rate parameters. The computer-generated reaction network, with associated rate expressions, is then converted to a set of mathematical equations, forming the kinetic model template. This template model can then be solved for different reactor systems within an optimization framework to tune the model with the process or experimental data. The product compositions can be calculated by solving the tuned model. With the established molecular structure-property correlations, the commercially relevant product properties can finally be evaluated. This automated molecule-based kinetic modeling strategy enables process chemists and engineers to focus on the fundamental chemistry and reaction kinetics at the molecular level and thus speed up the model development process.

1.5　THE PREMISE

Chemical engineering provides a rigorous framework for the construction, solution, and optimization of detailed kinetic models for delivery to process chemists and engineers. Relevant issues include the integration of the technical components of detailed kinetic modeling, namely, the modeling of reactant structures and compositions, the automated reaction network building and the use of the model building in "what if" scenarios, the organization of kinetic rate parameters, the solution of the kinetic model in the context of a reactor model, and the optimization of the model to experimental data, which are often in a vague and incomplete form, requiring assumptions and approximations for use. An overlying issue is that the delivery must be in a form that makes the model accessible to process chemists and engineers who may not be experts in computer hardware, operating systems, and programming languages.

This book is divided into two parts. Part I covers the development of tools for the construction, solution, and optimization of detailed kinetic models and the integration of the above-mentioned technical components. Chapter 2 addresses the molecular structure and composition modeling approach to convert complex feedstocks to a set of representative molecular structures. In Chapter 3, we exploit various techniques and methods to build and control the reaction network of complex process chemistries; various properties of the reaction network are also analyzed. In Chapter 4, the rate laws are discussed, and the linear free energy relationship (LFER) concepts are extended and generalized to organize and evaluate kinetic rate parameters of complex process chemistries. Chapter 5 discusses the mathematical background and model solving techniques for detailed kinetic models and proposes the model solving guidelines. In Chapter 6, these technical components of molecule-based detailed kinetic modeling are integrated into one complete system and a single user-friendly software package — the Kinetic

Modeler's Toolbox (KMT) — accessible on routine hardware and operating system combinations; various model delivery technologies are also discussed. Part II presents applications such as the verification of the developed modeling approach and KMT, including heavy naphtha reforming (Chapter 7), heavy paraffin hydrocracking (Chapter 8), naphtha hydrotreating (Chapter 9), gas oil hydroprocessing (Chapter 10), and gas oil FCC (Chapter 11), as well as naphtha pyrolysis (Chapter 12). Finally, in Chapter 13, the status is summarized with a view toward future work in the area of automated detailed kinetic modeling of complex processes.

REFERENCES

Aris, R., On reactions in continuous mixtures, *AIChE J.,* 35, 539–548, 1989.

Astarita, G. and Ocone, R., Lumping nonlinear kinetics, *AIChE J.,* 34, 1299, 1988.

Baltanas, M.A. and Froment, G.F., Computer generation of reaction networks and calculation of product distributions in the hydroisomerization and hydrocracking of paraffins on Pt-containing bifunctional catalysts, *Comput. Chem. Eng.,* 9, 71–81, 1985.

Bos, A.N.R, Lefferts, L., Marin, G.B., and Steijns, M.H.G.M., Kinetic research on heterogeneously catalysed processes: a questionnaire on the state-of-the-art in industry, *Appl. Catal. A: Gen.,* 160, 185–190, 1997.

Broadbelt, L.J., Stark, S.M., and Klein, M.T., Computer generated pyrolysis modeling: on-the-fly generation of species, reactions and rates, *Ind. Eng. Chem. Res.,* 33, 790–799, 1994.

Clymans, P.J. and Froment, G.F., Computer generation of reaction paths and rate equations in the thermal cracking of normal and branched paraffins, *Comput. Chem. Eng.,* 8, 137–142, 1984.

Dente, M., Ranzi, E., and Goossens, A.G., Detailed prediction of olefin yields from hydrocarbon pyrolysis through a fundamental simulation model (SPYRO), *Comput. Chem. Eng.,* 3, 61–75, 1979.

Froment, G.F., Fundamental Kinetic Modeling of Complex Refinery Process on Acid Catalysts, The Kurt Wohl Memorial Lecture, University of Delaware, 1999.

Gray, M.R., Lumped kinetics of structural groups: hydrotreating of heavy distillates, *Ind. Eng. Chem. Res.,* 25, 505–512, 1990.

Hillewaert, L.P., Dierickx, J.L., and Froment, G.F., Computer generation of reaction schemes and rate equations for thermal cracking, *AIChE J.,* 34(1), 17–24, 1988.

Jacob, S.M., Gross, B., Volts, S.E., and Weekman, V.W., A lumping and reaction scheme for catalytic cracking, *AIChE J.,* 22, 701–713, 1976.

John, T.M. and Wojciechowski, B.W., On identifying the primary and secondary products of the catalytic cracking of neutral distillates, *J. Catal.,* 37, 348, 1975.

Joshi, P.V, Molecular and Mechanistic Modeling of Complex Process Chemistries, Ph.D. Dissertation, University of Delaware, Newark, 1998.

Kmak, W.S., A Kinetic Simulation Model of the Powerforming Process, paper presented at AIChE Nat. Meet. Preprint, Houston, TX, 1971.

Laxminarasimhan, C.S., Verma, R.P., and Ramachandran, P.A., Continuous lumping model for simulation of hydrocracking, *AIChE J.,* 42(9) 2645–2653, 1996.

Liguras, D.K. and Allen, D.T., Structural models for catalytic cracking. 1. Model compounds reactions, *Ind. Eng. Chem. Res.,* 28(6), 665–673, 1989a.

Liguras, D.K. and Allen, D.T., Structure models for catalytic cracking. 2. Reactions of simulated oil mixtures, *Ind. Eng. Chem. Res.*, 28(6), 674–683, 1989b.

Marin, G.B. and Froment, G.F., Reforming of C6 hydrocarbons on a Pt-Al2O3 catalyst, *Chem. Eng. Sci.*, 37(5), 759–773, 1982.

Mudt, D.R., Hoffman, T.W., and Hendan, S.R., The Closed-Loop Optimization of a Semi-Regenerative Catalytic Reforming Process, AIChE paper-w51, 1995.

Qader, S.A. and Hill, G.R., Hydrocracking of Gas Oil, *I&EC Proc. Des., Dev.* 8(1), 98, 1969.

Quann, R.J. and Jaffe, S.B., Structure oriented lumping: describing the chemistry of complex hydrocarbon mixtures, *Ind. Eng. Chem. Res.*, 31(11), 2483–2497, 1992.

Quann, R.J. and Jaffe, S.B., Building useful models of complex reaction systems in petroleum refining, *Chem. Eng. Sci.*, 51(10), 1615, 1996.

Ramage, M.P., Graziani, K.R., Schipper, P.H., Krambeck, F.J., and Choi, B.C., KINPTR (Mobil's Kinetic Reforming Model): a review of Mobil's industrial process modeling philosophy, *Adv. Chem. Eng.*, 13, 193, 1987.

Stangeland, B.E., A kinetic model for the prediction of hydrocracker yields, *I&EC Proc. Des. Dev.*, 13, 71, 1974.

Sundaram, K.M. and Froment, G.F., Modeling of thermal cracking kinetics, *Chem. Eng. Sci.*, 32, 601–608, 1977a.

Sundaram, K.M. and Froment, G.F., Modeling of thermal cracking kinetics II, *Chem. Eng. Sci.*, 32, 609–617, 1977b.

Sundaram, K.M. and Froment, G.F., Modeling of thermal kinetics. 3. Radical mechanisms for the pyrolysis of simple paraffins, olefins and their mixtures, *Ind. Eng. Chem. Fund.*, 17, 174–182, 1978.

Van Damme, P.S., Narayanan, S., and Froment, G.F., Thermal cracking of propane and propane-propylene mixtures: pilot plant versus industrial data, *AIChE J.*, 21, 1065–1073, 1975.

Weekman, V.W., Lumps, models, and kinetics in practice, *AICHE Monograph Series*, 75, 1979.

Part I

Methods

2 Molecular Structure and Composition Modeling of Complex Feedstocks

2.1 INTRODUCTION

The first step in the conceptual development of a detailed molecule-based model for a complex feedstock is to determine an accurate molecular representation of the feedstock. The conventional analytical techniques usually cannot directly measure the identities of all the molecules in the complex feedstock, especially the high carbon number range, but only the indirect characteristics.

What can we do when we cannot measure? A concise approach to deal with the complexity of such a problem is to represent the molecules statistically. The overall steps of such an approach are presented in Figure 2.1, developed by Campbell (1998). The goal is to transform indirect analytical information about the molecules in a feedstock into a molecular representation. Both the identities and weight fractions of the molecules are sought. The former dictates the molecular structure, whereas the latter provides the quantitative initial conditions for a detailed molecule-based model.

The idea behind this stochastic modeling approach is as follows. Any molecule in a petroleum feedstock can be viewed as a collection of molecular attributes (number of aromatic rings, number of naphthenic rings, number of side chains, length of side chains, etc.). Neurock et al. (1990) developed a Monte Carlo construction technique whereby petroleum molecules are stochastically constructed by random sampling of probability distribution functions (PDFs), one for each molecular attribute. The PDF provides the quantitative probability of finding the value or less of a given attribute. Monte Carlo sampling of the set of PDFs provides a large ensemble of "computer" molecules whose properties can be compared to experimentally measured values.

Handling this large ensemble of computer molecules offers many challenges. In a rigorous deterministic molecule-based model, a mass balance differential equation is needed for each reactant and product species. The large number of reactant and product molecules then requires a very large system of equations, which will be too difficult to solve even in state-of-the-art computers. In order to develop an efficient model, however, the number of input molecules must

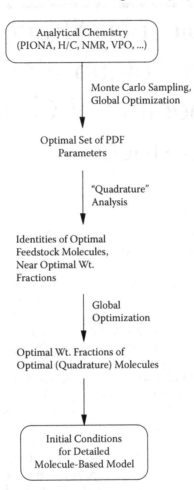

FIGURE 2.1 Flow diagram of stochastic modeling of molecular structures and compositions of a complex feedstock. (Rectangles indicate ... I/O[Input/Output].)

generally be limited to the order of 10^2 molecules, which is much smaller than the 10^4 molecules necessarily generated by Monte Carlo construction techniques (Petti, et al., 1994). To this end, a quadrature molecular sampling technique (Campbell, 1998) has been developed that generates a small number of quadrature molecules representative of a feedstock. These representative molecules are optimized to ensure that the small representation matches accurately the initial feedstock characterization. It is noteworthy that this small set of molecules can often match the characterization of the initial feedstock as well or even better than a much larger stochastic representation.

In this chapter, a brief review of the complex feedstock characterization techniques is first given, with an emphasis on the information that has been applied to the determination of an accurate molecular representation. This is followed by a detailed overview of the statistical representation and quadrature sampling of complex feedstocks. This structure modeling approach is then applied to a light gas oil case study. Finally, various issues concerning the molecular structure modeling of complex feedstocks are discussed in detail.

2.2 ANALYTICAL CHARACTERIZATION OF COMPLEX FEEDSTOCKS

State-of-the-art analytical techniques, such as the detailed hydrocarbon analysis (DHA) developed at Hewlett-Packard, have identified the individual hydrocarbon molecules of light petroleum fractions such as naphtha. Beyond C10, however, the number of possible isomers precludes a direct identification.

In general, the currently used typical analytical techniques will not provide the identities and concentrations of the molecules of a complex feedstock beyond the naphtha range, but rather indirect structural characteristics of the molecules. In order to construct a molecular representation, then, it is necessary to gather clues from these analytical tests as to the true molecular identities within the mixture.

Many analytical techniques have been used to probe the structure of complex feedstocks, although not all are equally useful to determine a molecular representation. Table 2.1 summarizes some common analytical methods that have been considered in our work to elucidate the molecular structures in complex feedstocks. The goal of each of these tests is to determine molecularly significant information that can be used when constructing a set of representative molecules. For some tests, the information will be directly applicable on

TABLE 2.1
Common Analytical Methods Used to Elucidate Molecular Structures in Complex Feedstocks

Characteristics	Analytical Methods
H/C ratio	Elementary analysis
Boiling point (BP)	Distillation
	Gas chromatography (GC)–simulated distillation (SimDis)
	Gas chromatography–mass spectrometry (GC-MS)
Compound class	High performance liquid chromatography (HPLC)
	PIONA (paraffin, isoparaffin, olefin, naphthene, aromatic)
	SARA (saturates, aromatics, resins, asphaltenes)
Molecular weight	Vapor pressure osmometry (VPO)
	Cryoscopy
	Gel permeation chromatography (GPC)
	Field ionization mass spectrometry (FIMS)
Atomic connectivity	^{1}H-, ^{13}C-NMR

a molecular basis by direct counting (e.g., hydrogen and carbon content). Other tests will be applicable on a molecule-by-molecule basis, although the prediction of the property for that molecule will be made by use of a simulation or correlation (e.g., boiling point). Also, some tests measure a bulk property of the feedstock for which a prediction can be made first by calculating the individual molecule's properties and then calculating the bulk property via a mixing rule (e.g., viscosity). A much more complete overview can be found in Campbell's work (1998).

Each test offers some insights either directly or indirectly into the structure of a complex mixture. The quality of this information may be affected by the precision of the analytical measurement. Furthermore, it is necessary to choose only a small number of analytical techniques to characterize a feedstock quickly, economically, and accurately. The goal in selecting an appropriate characterization is to determine enough structural detail with a desired level of precision that an accurate molecular representation may be constructed while also meeting any time or cost constraints.

2.3   MOLECULAR STRUCTURE MODELING: A STOCHASTIC APPROACH

As discussed in the previous section, a number of techniques can be used to gain valuable insights into the structure of complex feedstocks. However, transforming that information into an accurate molecular representation poses many challenges.

A statistical view of the complex feedstock provides a path forward. Any molecule in a petroleum feedstock can be viewed as a combination of structural attributes (number of aromatic rings, number of naphthenic rings, number of alkyl side chains, length of side chains, etc.), each of which is represented by a PDF. The PDF is a function that provides the probability of finding the value or less of a given attribute. By sampling the attribute PDFs, the values of the structural attributes for an individual molecule can be determined, which in turn specifies the molecule.

An illustration of the sampling technique is depicted in Figure 2.2. For each molecule, a random number (RN) is selected to first determine the molecule type (aromatic, naphthene, paraffin, olefin, etc.). Then a random number is selected for each attribute necessary to specify the molecule. For example, in the case of a naphthenic molecule, random numbers would be generated for attributes corresponding to the number of naphthenic rings, the number of alkyl side chains, and the length of side chains. In each case, the random number is compared to a PDF to determine the numerical value of the attribute.

Defining the appropriate molecular attributes and developing a construction algorithm are important aspects of this modeling technique. Once the molecules are represented by suitable structural attributes, the PDFs corresponding to these molecular attributes need to be optimized to match experimental data on the feedstock. The concepts of molecular attributes, PDFs, construction algorithms, and optimization are discussed in detail in the following subsections.

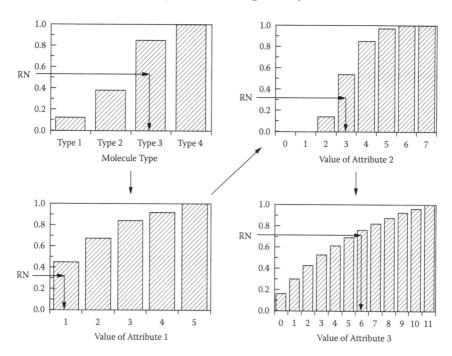

FIGURE 2.2 Illustration of the statistical sampling technique for determining molecule identity.

2.3.1 PROBABILITY DENSITY FUNCTIONS (PDFs)

In order to understand how to statistically represent a feedstock, it is necessary to first understand the concept of a PDF. A PDF may either be discrete (integral values of x only) or continuous (any real value of x). A discrete PDF is defined by the following equations:

$$0 \leq f(x) \leq 1 \tag{2.1}$$

$$\sum f(x) = 1 \tag{2.2}$$

A continuous PDF is defined similarly, except that Equation 2.3 is applicable instead of Equation 2.2:

$$\int_{-\infty}^{\infty} f(x)dx = 1 \tag{2.3}$$

TABLE 2.2
Probability Density Functions Used to Model the Structural Attributes of a Complex Feedstock

$$p_i = f(x_i, \alpha, \beta, \gamma)$$

$$x_i = \text{attribute}$$

$$\alpha, \beta, \gamma = pdf \text{ parameters}$$

Exponential:

$$p_i = \frac{e^{\left(-\frac{x_i - \gamma}{\Theta}\right)}}{\Theta}$$

2 parameters (γ, Θ)
$\gamma \leq x_i$
$\Theta = \mu - \gamma$
$\sigma = \mu - \gamma$
$\gamma = \text{minimum}$

Gamma:

$$p_i = \frac{\left((x_i - \gamma)^{(\alpha-1)} * e^{\left(-(x_i-\gamma)/\Theta\right)}\right)}{(\Gamma(\alpha) * \Theta^\alpha)}$$

3 parameters (γ, α, Θ)
$\gamma \leq x_i$
$\Theta = \sigma^2 / (\mu - \gamma)$
$\alpha = (\mu - \gamma)^2 / \sigma^2$
$\gamma = \text{minimum}$

Chi-square:

$$p_i = \frac{\left((x_i - \gamma)^{\left(\left(\frac{r}{2}\right)-1\right)} * e^{\left(-(x_i-\gamma)/2\right)}\right)}{\left(\Gamma\left(\frac{r}{2}\right) * 2^{r/2}\right)}$$

2 parameters (γ, r)
$\gamma \leq x_i$
$r = \mu - \gamma$
$\alpha = (2(\mu - \gamma))^{0.5}$
$\gamma = \text{minimum}$

Common discrete PDFs include the discrete uniform distribution, the binomial distribution, and the Poisson distribution. Examples of continuous distributions include the normal distribution, the gamma distribution, and the exponential distribution. Functional forms of some common PDFs are shown in Table 2.2.

In addition to knowing the rigorous definition of a probability density function, there are several other important practical issues that need to be considered when using them to construct a complex feedstock. It is important to know whether using such an approach has some physical meaning attached to it. Deciding on an appropriate functional form of the distributions that are used to model the feedstock is important as well. Finally, issues in discretizing and truncating distributions and conditional probability must be considered for the accurate representation of a complex feedstock.

2.3.1.1 PDFs Used to Describe Complex Mixtures

The concept of using PDFs to describe complex mixtures has existed for a long time. Flory (1936) developed a modified gamma distribution to describe the molecular size distribution of condensation polymers. Libanati (1992) studied the thermal degradation of an infinite polymer and indicated that the molecular weights of the products followed a log-normal distribution. The application of probability distribution functions was extended to petroleum fractions when it was hypothesized and later confirmed that kerogen, which breaks down to form oil, could be modeled as an infinite polymer, and thus, the molecular weight distribution of oil should be similar to that of polymer products. Shibata et al. (1987) used mixed distributions to enhance phase equilibrium calculations for a petroleum reservoir. Light molecules (methane, ethane, etc.) were represented as discrete components, while the C_7+ fraction was described in terms of a continuous distribution (exponential, gamma, and normal distributions were discussed). Whitson (1990) used a gamma distribution (which is similar in shape to a log-normal distribution) to fit the molar and weight distribution of the C_7+ fraction of crude oil, further supporting the notion of representing crude components with PDFs.

In addition to earlier modeling efforts, there is direct experimental evidence that statistical distributions can be used to model petroleum. Pederson et al. (1992) used high-temperature gas chromatography to measure the weight percent distribution of carbon number up to $C_{80}+$ for 17 North Sea oils. They showed that an exponential distribution fit to C_{20} could be used to accurately predict the quantities of heavier components. It should be noted that n-paraffin standards were used to correlate retention time to carbon number. Boduszynski (1987) has pointed out the wide divergence in boiling points with increasing carbon number for different compound classes. Thus, a more accurate description of results would be that the weight percent distribution of boiling point was fit using an exponential distribution. By fitting the boiling point distribution up to the boiling point of C_{20} (344°C), the rest of the distribution could be predicted.

Petti et al. (1994) and Trauth et al. (1994) extended the use of PDFs to model not only the molecular weights and boiling points, but also the structural attributes described in the previous section. Experimental proof that such an approach is valid was provided by a statistical modeling project that dealt with the thermal depolymerization of coal (Darivakis et al., 1990). As with the degradation of an infinite polymer, a gamma distribution accurately fit the molecular weight distribution of the products. Since products are formed primarily by bond fission reactions during pyrolysis, this result indicates that the individual structural attributes also would be well represented by gamma distributions. Trauth (1990) demonstrated that using a gamma distribution for each of the structural attributes of a petroleum resid yielded a molecular weight distribution that could also be represented by a gamma distribution. By optimizing the PDF parameters so that a stochastically determined molecular representation closely matched a set of analytical characterizations, it was shown that many of the key properties of the resid could be simulated.

2.3.1.2 Molecular Structural Attributes

In order to construct a molecular representation, it is necessary to first identify the "building blocks" of a molecule. On the most basic level, a molecule is defined by a juxtaposition of atoms that are chemically bonded together in some specific manner. In principle, a molecule can be constructed by randomly choosing and connecting atoms. However, not all atoms can be chosen independently. For instance, if an aromatic carbon is first selected, enough other aromatic carbons must now be chosen to complete the aromatic structure. This group of six aromatic carbon atoms that complete the aromatic ring is now defined as an irreducible structural group.

The structural attribute that is related to the irreducible structural group is defined to account for this. A structural attribute is defined as an element of structure that is represented by a PDF. This is different from the irreducible structural group because some irreducible structural groups are defined by multiple attributes. For instance, to specify a molecule, both the number and length of alkyl side chains must be specified; however, the irreducible structural group is the alkyl side chain. Therefore, toluene would have two irreducible structural groups: an aromatic ring and an alkyl side chain. To specify a toluene molecule, however, requires three attributes: one aromatic ring, one alkyl side chain, and one carbon in the alkyl side chain. Similarly, for more complex molecules, the configuration of the rings and placement of the side chains must be specified as molecular attributes.

2.3.1.3 Appropriate PDF Forms

As mentioned in Section 2.3, PDFs have many functional forms. Selecting the appropriate form can be very important for optimally representing a feedstock. The most important consideration is that the PDF qualitatively captures the shape of what is being modeled. In addition, it is often desirable to select distributions that are flexible, so that slight deviations from a particular functional form can be accurately modeled. It may also be desirable to minimize the number of parameters that must be optimized, particularly when many distributions have to be used to model a feedstock.

Many principles used to model molecular weight or boiling point distributions can be used to give insight into an acceptable strategy for modeling compounds at an attribute level. The molecular weight or boiling point distributions generally follow a smooth curve for petroleum fractions. Figure 2.3 shows the relative boiling point distributions of a petroleum kerosene and a petroleum resid. Lighter fractions like those for kerosene are characterized by both a minimum and a maximum boiling point and generally have a boiling point distribution that is normal or skewed normal. A petroleum resid is defined only by a minimum boiling point. Generally, a petroleum resid boiling point distribution is characterized by a rapid rise followed by a slow decrease that mathematically would be characterized by a gamma type distribution. Investigations of both polymers and heavy components of fossil fuels indicate that functional forms like gamma distributions or exponential distributions accurately model such systems.

The boiling point distribution is closely linked to the structure of the molecules. Generally, boiling point distributions are very closely correlated to the molecular weight or carbon number of a species. Furthermore, the molecular

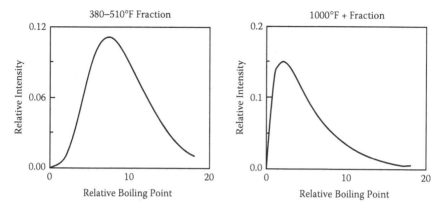

FIGURE 2.3 Relative boiling point intensity for kerosene and vacuum resid petroleum fractions.

attributes described in the previous section are implicitly related to the carbon number, so the attribute PDFs should be well modeled by the same type of distribution that models the boiling point distribution.

The foregoing semi-theoretical arguments are reinforced by empirical experience. Trauth et al. (1994) showed that a set of experimentally determined analytical properties of a petroleum resid could be well represented by modeling the structural attributes with gamma and gamma-like distributions. Furthermore, the gamma distribution ranges from an exponential distribution to a delta function and can also approximate a normal distribution. This flexibility allows for the modeling of lighter feedstocks as well, even though the boiling point distributions for such feedstocks may not be considered typically gamma.

A final consideration in choosing a functional form is the number of parameters that must be specified. Although the gamma distribution is quite flexible, it also requires three parameters. Trauth et al. (1994) the functional forms shown in Table 2.2 to model a series of petroleum resids. These can be seen graphically in Figure 2.4. The chi-square distribution is a special case of the gamma distribution where the standard deviation equals half of the mean. The gamma distribution can also match the exponential distribution for certain values of parameters. What is gained by using a chi-square or exponential distribution is one fewer parameter that needs to be optimized. This is important and will be discussed in further detail with CPU time requirements.

2.3.1.4 Discretization, Truncation, and Renormalization

Although PDFs such as the gamma distribution and the exponential distribution can be used to model complex feedstocks accurately, both of these distributions are continuous. However, real feedstocks are composed of attributes with discrete integer values. Therefore, it is necessary to transform these continuous distributions

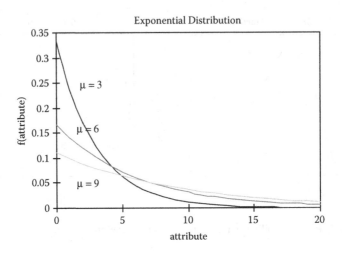

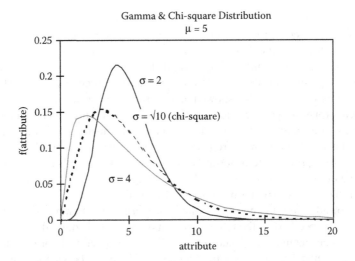

FIGURE 2.4 Examples of exponential, gamma, and chi-square distributions.

into discrete distributions. In order to transform a continuous distribution into a discrete distribution, it is necessary to divide the distribution into intervals and evaluate a representative value for each interval. Trauth (1990) showed that the shape of a continuous gamma curve was well maintained using such a discretization.

Just as the attributes of the molecules in a complex feedstock are all discrete integer values, they are also finite. Therefore, it is necessary to truncate the distributions at some physically reasonable value. Once a distribution has been truncated, it is also necessary to renormalize the distribution so that the probabilities add up to 1. A truncation criterion may be set by specifying the minimum contribution each new interval must make to the cumulative distribution expressed on a fractional basis. Trauth (1990) found that a value of 0.01 worked well.

2.3.1.5 Conditional Probability

One final important consideration for generating a set of PDFs is conditional probability. Conditional probability can be defined as the constraint one attribute value realization might have on the PDF used to obtain another. In other words, once an attribute value has been determined, the probability of another attribute having a particular value may be altered.

Consider, for example, the case of a petroleum resid that is defined only by a minimum boiling point. Knowing that every molecule has to boil above that minimum boiling point requires that there be some conditional probability. For instance, a three-ring aromatic compound would need more side chain carbons to boil above 1000°F than would a four-ring aromatic compound. Therefore, the probability distribution for the number of alkyl side chains or the length of the alkyl side chains must be different for the two different attribute values of number of aromatic rings.

The implementation of the accounting for these types of physical constraints is very important in obtaining a good molecular representation. In addition to initial boiling point criteria, it is also necessary to incorporate conditional probability to limit the size of a molecule or to limit the number of heteroatoms in a molecule.

2.3.2 MONTE CARLO CONSTRUCTION

The previous sections presented the idea of a molecular attribute, which is modeled by a PDF. In addition to this basic definition of the molecules, it is necessary to know how to select attribute values from the PDFs in order to build a sample of representative molecules.

One attractive technique is Monte Carlo sampling, where random numbers are matched to the PDFs and the combinations of the attributes represent the molecules. The idea of stochastic sampling ensures that the molecular representation is not biased. This is especially important since the number of molecules that will be generated will be much less than the number of possible molecules.

2.3.2.1 Monte Carlo Sampling Protocol

A representative set of molecules can be built for a given feedstock by stochastic sampling of the molecular attribute PDFs. The complex feedstocks, however, often consist of molecules with multiple attributes, each of which is defined by a PDF. Therefore, it is necessary to determine the order in which attributes must be sampled.

Before a molecule can be specified, the type of molecule that is being constructed must be determined. Therefore, the first PDF that is sampled will be the compound class distribution. For instance, when building a petroleum gas oil molecule, one would first determine whether the molecule is a paraffin, an iso-paraffin, a naphthenic, or an aromatic.

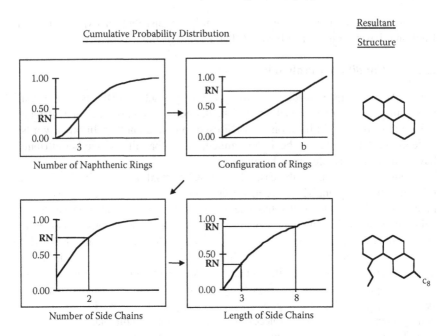

FIGURE 2.5 Stochastic construction of a naphthenic molecule.

Once a molecule type has been determined, the PDFs that define the attributes for that molecule can, in principle, be sampled in any order to specify the molecular structure. However, since certain attribute values may be dependent on other attribute values, as discussed in Section 2.3.1.5, ordering the sampling of the PDFs may be beneficial.

Figure 2.5 shows how a naphthenic molecule is constructed. The number of naphthenic rings is determined first, since this will have an effect on the number and length of alkyl side chains. Once the naphthenic core is determined, the number of side chain carbons that must be present to ensure that the molecule boils within a proper range can be determined, and the PDFs for the number and length of alkyl side chains can be adjusted accordingly.

2.3.2.2 Optimal Representation of a Complex Feedstock

In addition to knowing how to construct a molecular representation for a complex feedstock, it is important to evaluate how well a given set of molecules represents a feedstock. In particular, a molecular representation should retain the key analytical properties that were experimentally determined for the feedstock. In addition to the desired accuracy of a model, it is also important to build the molecules in a time-efficient manner.

With the molecular information from various analytical and reaction techniques in hand, the goal becomes constraining the molecular representation to these analytical data. The chi-square statistic is a typical objective function used to optimize a representative feed to an actual feed:

$$
\begin{aligned}
2 = &\left(\frac{MW_{exp} - MW_{pred}}{0.05 * MW_{exp}} \right)^2 + \left(\frac{HtoC_{exp} - HtoC_{pred}}{0.02 * HtoC_{pred}} \right)^2 + \\
&+ \left(\frac{Halpha_{exp} - Halpha_{pred}}{0.02} \right)^2 + \left(\frac{Harom_{exp} - Harom_{pred}}{0.01} \right)^2 \\
&+ \left(\frac{1}{\#Comps} \right) \sum_{i=1}^{\#Comps} \left(\frac{PIONAWt_{i,exp} - PIONAWt_{i,pred}}{0.03} \right)^2 \\
&+ \left(\frac{1}{\#Fracs} \right) \sum_{i=1}^{\#Fracs} \left(\frac{SIMDIS_{i,exp} - SIMDIS_{i,pred}}{0.01} \right)^2 \\
&+ \left(\frac{1}{\#Comps} \right) \sum_{t=0}^{T} \sum_{i=1}^{\#Comps} \left(\frac{Wtfrac_{i,exp} - Wtfrac_{i,pred}}{0.01} \right)^2 + ...
\end{aligned}
\tag{2.4}
$$

The numerator is the square of the difference between the model prediction (i.e., computer molecules' properties) and the experimentally determined properties. The denominator is a weighting factor equal to the standard deviation of the experimentally determined value. This objective function can be modified easily for any particular analytical information on a feed. For instance, the PIONA (paraffin, isoparaffin, olefin, naphthene, aromatic) weight fractions shown in Equation 2.4 were used in the objective function for lighter fractions. For heavier fractions, such as resid, SARA (saturates, aromatics, resins, asphaltenes) separations are generally used instead. If other data, such as ^{13}C NMR, are available, other terms can easily be added onto this flexible objective function as needed.

All the terms except the last one in Equation 2.4 represent the analytical characterization of the feedstock, as discussed in Section 2.2. Vasudeva (1999) has added the last term, a reaction term, to the objective function to utilize the reactivity and product information to optimize the molecular representation of the feedstock. If the goal is to determine the molecular structures in the feedstock and a good reaction model is already in hand, using the reactivity information will certainly help to better determine the feedstock structures. However, if the goal is to develop good reaction models, a more comprehensive optimization scheme including optimizing both PDF parameters and rate parameters at the same time can be considered, which will be discussed in Chapter 6.

The above objective function can be used in the context of a global optimization routine where the PDF parameters are varied, and a representative set of molecules is built for each case. These computer molecules can then be compared to the real feedstock, using Equation 2.4 to determine how well a given set of molecules determined from a particular set of PDFs matches the experimental measurements. Lower values of the objective function indicate that a molecular representation matches the experimental data better.

2.3.2.3 Sample Size

The number of sample molecules selected must be small enough to allow for time-efficient CPU optimization. However, if the sample size is too small, the large variation in the objective function may make different sets of PDF parameters statistically indistinguishable, even though larger samples would allow for differentiation.

Petti et al. (1994) investigated the optimal sample size needed to represent a petroleum resid. Figure 2.6 is a reproduction of a graph from his paper. The x-axis indicates the number of molecules that are being constructed. The left y-axis is the value of the objective function, and the right y-axis is the amount of CPU seconds needed to construct the representation. The dashed line indicates the mean

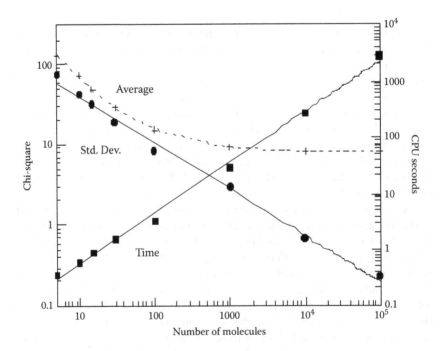

FIGURE 2.6 Chi-square value and CPU time requirements versus the number of molecules constructed for a petroleum resid (Petti et al., 1994). See text for details.

value of the objective function for the sample, and the solid line with the negative slope is a measure of the standard deviation of the objective function. The solid line with a positive slope is the amount of CPU time required for construction.

Figure 2.6 shows that increasing the sample size decreases the variability. Therefore, as large a sample size as possible is desired. In addition, the mean value of the objective function decreases until approximately 1000 molecules are constructed. This means at least 1000 molecules must be built for an accurate measure of the objective function.

The downside to increasing the number of molecules for each representation is shown by the CPU seconds versus the number of molecules. Construction of 100,000 molecules requires nearly 20 minutes. This amount of time is magnified by the global optimization process, where hundreds or thousands of representative samples will be built when determining the optimal parameter values.

Therefore, a balance between the desired accuracy and the time requirements must be established. Petti et al. (1994) determined that a sample size on the order of 10,000 optimally balances the CPU demand and the desired accuracy of the objective function.

2.3.3 QUADRATURE MOLECULAR SAMPLING

Irrespective of the details that might be specific for each feedstock, an optimal set of PDFs will allow Monte Carlo construction of a representation of the feed in terms of thousands of molecules. This number of initial molecules will often be too large for reaction modeling purposes. For example, the reaction model will often be the kernel of a larger process such as the computational fluid dynamics (CFD) model, which, in turn, may be used for a larger scale of optimization or control. Thus, there will often be considerable motivation to construct a smaller, less CPU-demanding model. To fix ideas, consider the case where it is desirable to represent the feedstock information in terms of about 10 to 100 molecular species. This requires an ordered sampling technique of the attribute PDFs that preserves the information within the PDFs.

2.3.3.1 Quadrature Sampling Protocol

As with the Monte Carlo construction of a molecular representation, it is necessary to determine how many molecules must be built and in what way the PDFs will be sampled. This means that both the number of values for each attribute and the combinations of attributes to be used must be chosen. Furthermore, the values must be selected in such a way that the molecular representation matches the key analytical properties needed for modeling purposes.

Experimental design protocols assist in establishing an acceptable number of species. For example, a full two-level factorial design would sample each PDF at two levels (high and low), and the resulting molecules would be combinations of these attribute values. Since the naphthenic fraction is determined by three attributes, a two-level factorial would result in eight representative naphthenic molecules. In practice, the objective function used to optimize the PDFs is not equally sensitive to each attribute, so it is often desirable to use different numbers of nodes (attribute values).

Campbell (1998) investigated the sensitivity of the objective function shown in Equation 2.4 to various attributes, which led to conclusions about the relative number of nodes that should be used to represent each PDF. For heavier petroleum fractions like resid, the most sensitive attributes were generally the number of aromatic rings and the number of naphthenic rings. Both the length of side chains and the paraffin length generally affected the objective function only slightly. For lighter fractions like naphtha, the paraffin length and number of main chain and side chain carbons affects the objective functions the most.

With the number of nodes for each PDF determined, it was then necessary to establish a method to determine the representative attribute values. One possibility is to use numerical quadrature. The integral of a continuous distribution function $p(x)$ can be solved using n quadrature points according to the following equation:

$$\int_x p(x)dx = \sum_1^n w(x_p)p(x_p) \tag{2.5}$$

In this equation, x_p corresponds to the quadrature point, $p(x_p)$ is the function $p(x)$ evaluated at the quadrature point, and $w(x_p)$ is the weighting factor. The values of the nodes (x_p) and the weighting factors can be determined by various methods. Typically, a Gauss–Laguerre quadrature is used for a function with a form of e^{-x}, and a Gauss–Hermitian is used for a function with a form of e^{-x^2}.

Preliminary investigations attempted to use the quadrature nodes of the gamma distributions. A Gauss–Laguerre quadrature method was first used to determine the nodes (attribute values) and weights (mole fractions) on the gamma distribution. This often resulted in quadrature nodes that were too heavily distributed on the lower end of the PDFs. Gauss–Hermitian quadrature was also investigated for those gamma distributions that closely approximated normal distributions. However, the quadrature points still failed to span the distributions adequately.

The goal in determining the attribute values that will represent the feedstock is to ensure that the attribute PDFs are well represented. One appealing method chooses attribute values based on equiprobable regions. For instance, if a PDF is to be represented by two nodes, the probability space is divided into two regions (0 to 0.5 and 0.5 to 1). If the attribute values for two adjacent regions coincide, the relative concentration of this attribute value is doubled. The representative values of PDFs are the attribute values associated with the midpoints of the probability regions. This method ensures that the PDFs are sampled over a wide range. Furthermore, since the attributes now represent equiprobable regions, the mole fractions of a particular class of molecule will be equal. From a mathematical viewpoint, this is equivalent to estimating the integrand as a sum of rectangles. Although this simple quadrature method is not as mathematically efficient as other quadrature methods, it ensures that the PDFs can be well spanned.

2.3.3.2 Fine-Tuning the Quadrature Molecular Representation

This process of representing the information in the PDFs in terms of a finite number of molecules represents, in principle, a loss of information. This suggests that further fine-tuning of the molecular representation would be desirable. Since, at this point, the structures of the molecules are fixed, only the mole fractions remain to be adjusted. This fine-tuning optimization adjusts the mole fractions to obtain an optimal match with the original characterization data using the same objective function used in developing the optimal PDFs. Previous work on resid and naphtha (Campbell, 1998) has shown that this fine-tuning of the mole fractions of even a small fixed set of molecules (~10 to 100) can accurately reproduce the properties of a wide variety of resid molecules.

2.4 A CASE STUDY: LIGHT GAS OIL

The molecular structure modeling approach discussed in the previous section was applied to light gas oil for its validation. The analytical properties of a typical type of light gas oil A are summarized in Table 2.3. The average molecular weight, H to C ratio, PINA (paraffin, isoparaffin, naphthenic, and aromatic) distribution, and simulated distillation results were used to optimize the PDF parameters for this feedstock.

TABLE 2.3
Analytical Properties of Light Gas Oil A

	Light Gas Oil
Molecular weight	181.4
H to C ratio	1.67
PINA fractions (wt. %)	
Paraffin	4.4
Isoparaffin	9.9
Naphthenic	32.0
Aromatic	53.7
Simulated Distillation	
10% cutoff	211.4°C
30% cutoff	248.0°C
50% cutoff	271.0°C
70% cutoff	295.6°C
90% cutoff	331.7°C
Final boiling point	403.5°C

The molecules were deconvoluted into their structural attributes, and each attribute was modeled by a probability density function. For this light gas oil, the naphthenic compounds were considered to have up to three rings; the aromatic compounds were considered to have two aromatic rings or one aromatic with up to two naphthenic rings. The following structural attributes were identified: normal paraffin length, isoparaffin carbon number, and number of carbons in each kind of ring compound (one-, two-, three-ring naphthenes; one-, two-ring aromatics; and two-, three-ring hydroaromatics).

Since gamma distribution is very flexible and can approximate many other distributions (e.g., exponential, normal, delta), it was chosen to model each structural attribute for the gas oil structure modeling.

The construction algorithm for the light gas oil is shown in Figure 2.7. The compound class of the molecule must first be determined. The order for specifying the structural attributes of a molecule is then designed to take advantage of the conditional probability considerations.

The PDF parameters for this light gas oil were optimized using a global simulated annealing technique. This was accomplished through an iteration of building an ensemble of 100,000 molecules via Monte Carlo sampling of the PDFs, comparison of the ensemble properties with the experimental values, and adjusting the PDF parameters until an objective function was minimized. The optimized PDF parameters are summarized in Table 2.4. Any structural feature

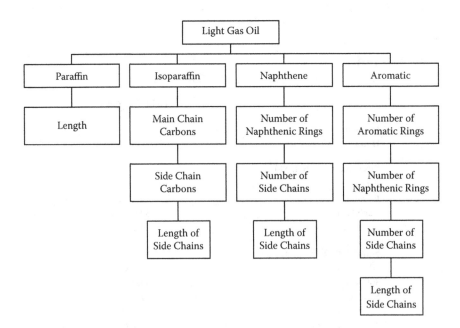

FIGURE 2.7 Construction algorithm for light gas oil.

TABLE 2.4
Optimal PDF Parameters for Light Gas Oil A

Structural Attribute	Light Gas Oil
Paraffin length	14.73(3.74)
Isoparaffin number of carbons	13.89(2.01)
One-ring naphthenics carbon number	14.64(5.08)
Two-ring naphthenics carbon number	16.57(0.77)
Three-ring naphthenics carbon number	18.92(0.42)
One-ring aromatics carbon number	13.77(1.22)
Two-ring aromatics carbon number	12.98(1.96)
Two-ring hydroaromatics carbon number	14.98(3.48)
Three-ring hydroaromatics carbon number	17.03(1.49)

not listed in Table 2.4, such as location of a branch, is assumed to follow a uniform distribution.

The quadrature molecules are list in Table 2.5, which shows the optimal molecular representation of the light gas oil as characterized in Table 2.3. The model predictions are shown in Table 2.6. For the most part, the model predictions are good, considering the very limited characterization information supplied.

TABLE 2.5
Optimal Quadrature Molecules for Light Gas Oil A

Molecule	Mole Fraction	Molecule	Mole Fraction
Tridecane	0.016	Eicosane	0.017
6-Methyl tridecane	0.001	4-Ethyl dodecane	0.005
5,6-Dimethyl dodecane	0.011	3,5-Diethyl decane	0.004
2-Methyl tetradecane	0.003	6-Ethyl tridecane	0.009
4-Ethyl tetradecane	0.002	3,5-Dimethyl tetradecane	0.003
6,6-Diethyl dodecane	0.004	5-Methyl pentadecane	0.005
6-Ethyl tetradecane	0.009	4,6-Dimethyl tetradecane	0.021

C_8 ring structure	0.034	C_9 ring structure	0.026
C_{10} ring structure	0.015	C_{11} ring structure	0.009

(Continued)

TABLE 2.5 (Continued)
Optimal Quadrature Molecules for Light Gas Oil A

Molecule	Mole Fraction	Molecule	Mole Fraction
C_2 (decalin substituted)	0.005	C_3 (decalin substituted)	0.088
C_4 (decalin substituted)	0.010	C_5 (decalin substituted)	0.027
C_7 (decalin substituted)	0.074	C_3 (perhydroanthracene substituted)	0.002
C_4 (perhydroanthracene substituted)	0.006	C_5 (benzene substituted)	0.097
C_6 (benzene substituted)	0.029	C_2 (naphthalene substituted)	0.006
C_3 (naphthalene substituted)	0.010	C_4 (naphthalene substituted)	0.027
(indane)	0.037	(tetralin)	0.130
C_4 (tetralin substituted)	0.113	C_7 (tetralin substituted)	0.081
C_1 (fluorene-type substituted)	0.018	C_3 (fluorene-type substituted)	0.038
(fluorene-type)	0.007		

TABLE 2.6
Model Predictions for the Analytical Properties

	Experimental	Prediction
Molecular weight	181.4	186.6
H to C ratio	1.67	1.64
PINA fractions (wt. %)		
Paraffin	4.4	4.2
Isoparaffin	9.9	9.0
Naphthenic	32.0	32.9
Aromatic	53.7	53.9
Simulated Distillation		**% off**
10% cutoff	10.0	10.0
30% cutoff	30.0	29.9
50% cutoff	50.0	49.9
70% cutoff	70.0	70.0
90% cutoff	90.0	90.0
Final boiling point	100.0	100.0
Objective function		1.4

2.5 DISCUSSIONS AND SUMMARY

It is possible to develop a statistical representation of a complex feedstock by constraining the representation to a set of analytical characterizations. However, it is important to balance time with accuracy. Realistically representing a complex feedstock often requires the use of detailed distributions to define the structural attributes. Additionally, larger stochastic representations will give a more precise value of the objective function used to optimize the PDF parameters. Each of these issues, however, requires extra CPU time, so that minimizing these while maintaining the desired level of accuracy should be the primary objective.

A quadrature sampling method to transform the basic analytical chemistry for a complex feedstock into a small set of representative molecules has been developed. The molecules are defined as a collection of attributes, whose values are represented by a PDF. For a petroleum feedstock, gamma distributions represented the probability of the attribute values fairly well. The information in these PDFs can typically be preserved in a set of 10 to 100 molecules. The key analytical properties of the feed are preserved by a systematic sampling of the PDFs in a manner that spans the PDFs and by further optimizing the mole fractions of the resulting molecules in order to account for any information lost in the transformation.

The more analytical information is available, the better this molecular structure modeling approach works. Since the objective function used to optimize the PDF parameters and the quadrature mole fractions is very flexible to customize, it can accommodate any analytical information to improve the accuracy of the model prediction. Furthermore, if some molecular structure information is already known, it can be fixed, and only the unknown structure attributes are optimized. In addition, reaction modeling of these molecules may give further insight into the initial feedstock structure. It is possible to make use of the reaction product information through the optimization of a more comprehensive objective function.

Another important issue is that although an optimal molecular representation might consist of typical molecules in the complex feedstock, the identities of these molecules may vary from feed to feed. Such differences are especially expected for such small representations. Varying molar fractions within a fixed set of molecules for the same kind of feedstock (such as gas oil or resid) can generally account for average properties (such as feed molecular weight and H to C ratio), but finer detail (such as NMR data) may lead to different structures for different feeds. A way to obtain a generic molecular representation for each kind of feedstock is to sum all the optimal representative molecules together for a set of typical feeds by applying the modeling approach developed here; then only the molar fractions are needed to further optimize to match the analytical properties.

A molecular representation of a complex feedstock thus provides the input to a reaction network generator to carry out chemical reactions at the molecular level in a specific process chemistry, which will be discussed in detail in the next chapter.

REFERENCES

Boduszynski, M.M., Composition of heavy petroleums. 1. Molecular weight, hydrogen deficiency, and heteroatomic concentration as a function of atmospheric equivalent boiling point up to 1400°F, *Energy Fuels*, 1, 2, 1987.

Campbell, D.M., Stochastic Modeling of Structure and Reaction in Hydrocarbon Conversion, doctoral dissertation, University of Delaware, Newark, 1998.

Darivakis, G.S., Peters, W. A., and Howard, J. B., Rationalization for the molecular weight distribution of coal pyrolysis liquids, *AIChE J.*, 36(8), 1189, 1990.

Flory, P.J., Molecular size distribution in linear condensation polymers, *J. Am. Chem. Soc.*, 58, 1877, 1936.

Libanati, C., Monte Carlo Simulation of Complex Reactive Macromolecular Systems, Ph.D. dissertation, University of Delaware, newark, Newark, 1992.

Neurock, M. N., Nigam, A., Libanati, C., Klein, M. T., Monte Carlo simulation of complex reaction systems: molecular structure and reactivity in modelling heavy oils, *Chem. Eng. Sci.*, 45(8), 2083–2088, 1990.

Pederson, K.S., Blilie, A.L., and Meisingset, K.K., PVT calculations of petroleum reservoir fluids using measured and estimated compositional data for the plus fraction, *Ind. Eng. Chem. Res.*, 31, 1378–1384, 1992.

Petti, T.F., Trauth, D.M., Stark, S.M., Neurock, M.N., Yasar, M., and Klein, M.T., CPU issues in the representation of the molecular structure of petroleum resid through characterization, reaction, and Monte Carlo modeling, *Energy Fuels*, 8(3), 570–575, 1994a.

Shibata, S.K., Sandler, S.I., and Behrens, R.A., Phase equilibrium calculations for continuous and semicontinuous mixtures, *Chem. Eng. Sci.*, 42, 1977–1988, 1987.

Trauth, D.M., Structure of Complex Mixtures through Characterization, Reaction, and Modeling, Ph.D. dissertation, University of Delaware, Newark, 1990.

Trauth, D.M., Stark, S.M., Petti, T.F., Neurock, M.N., and Klein, M.T., Representation of the molecular structure of petroleum resid through characterization and Monte Carlo modeling, *Energy Fuels,* 8(3), 576–580, 1994.

Vasudeva, A., Stochastic Modeling of Heavy Petroleum Feedstock: Use of Reactivity Information in Structure Determination, MS thesis, University of Delaware, Newark, 1999.

Whitson, C.H., Characterizing hydrocarbon plus fractions, *Soc. Pet. Eng. J.,* 683, 1990.

3 Automated Reaction Network Construction of Complex Process Chemistries

3.1 INTRODUCTION

Many legacy process models were implemented using lumped kinetics schemes, where the molecules are grouped according to their global properties, such as boiling point fractions or solubilities. Molecular information is obscured by the multicomponent nature of each lump. However, increasing economic needs and environmental concerns have focused attention on the molecular composition of complex feedstocks and their refined products. Modeling approaches that allow for reaction of complex feedstocks and prediction of molecular properties require an unprecedented level of molecular detail.

Modern analytical measurements indicate the existence of at least $O(10^5)$ unique molecules in petroleum feedstocks. The sheer size of this modeling problem engenders a conflict between the need for molecular detail and the formulation of the model. In rigorous deterministic molecular models, the mass balance for each species is equivalent to one differential equation. The formulation, as well as the solution, of the implied model is therefore formidable. Thus, it is compelling to develop computer algorithms not only to solve but also to formulate the model, thereby allowing the researchers to focus on the fundamental chemistry and reaction rules.

The advancement of analytical techniques and computer technology has made the building and implementation of such detailed molecular models possible. To be able to automate the reaction network building process, the computer has to be taught how to recognize the molecules and carry out chemical reactions. Broadbelt et al. (1994a,b, 1996) used a graph-theoretic approach and developed such algorithms to automate the reaction model building for ethane pyrolysis. The main ideas of the graph-theory concepts of automated model building are summarized here.

A molecule can be represented by a graph, with the atoms being the nodes and the bonds being the edges. A more mathematically tractable implementation for a chemical species is through its bond–electron matrix, where the ij entries denote the bond order between connected atoms i and j, and the ii entries denote the number of nonbonded electrons or ions surrounding the atom i. Mechanistic

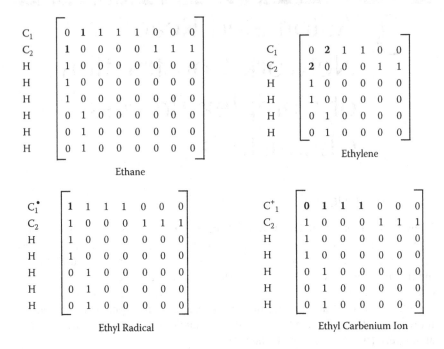

FIGURE 3.1 Bond–electron matrix representation of molecules, free radicals, and ions.

modeling of complex chemistries involves not only molecular species but also radical and ionic species as intermediates. The radical center is represented by an entry 1 in the diagonal position of the atom with the radical center. The subtraction of the sum of the entries in the row for a particular atom, from the valency of that atom, is assigned as the ionic charge on that particular atom. Figure 3.1 illustrates this idea with the connectivity matrix representation of some small (C2) molecules found in hydrocarbon systems, such as ethane, ethylene, ethyl radical, and ethyl ion.

This bond–electron matrix representation enables chemical reactions to be implemented through a simple matrix addition operation (Ugi et al., 1979). Addition of a reaction matrix, the entries of which denote the change in connectivity and change in electronic environment when a species undergoes a particular reaction, to the bond–electron reactant matrix yields a matrix representing reaction products. For all of the reactions in complex chemistries, the connectivities of only a few of the atoms in the involved molecules change during the reaction. Thus, the bond breakage and formation information can be summarized succinctly in terms of a formal reaction matrix for each reaction family. Figure 3.2 depicts a representative fission reaction of ethane into two methyl radicals.

The reaction network can then be constructed through repetitive application of reaction matrices to the reactants and their progeny. Figure 3.3 shows the

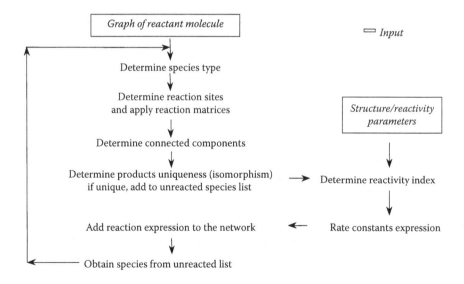

$$\begin{array}{c} C_1 \\ C_2 \end{array} \begin{vmatrix} 0 & 1 \\ 1 & 0 \end{vmatrix} \quad + \quad \begin{vmatrix} 1 & -1 \\ -1 & 1 \end{vmatrix} \quad = \quad \begin{vmatrix} 1 & 0 \\ 0 & 1 \end{vmatrix}$$

Reduced Reactant Matrix **Reaction Matrix** **Reduced Product Matrix**

Reactant Matrix

C_1	0	1	1	1	1	0	0	0
C_2	1	0	0	0	0	1	1	1
H	1	0	0	0	0	0	0	0
H	1	0	0	0	0	0	0	0
H	1	0	0	0	0	0	0	0
H	0	1	0	0	0	0	0	0
H	0	1	0	0	0	0	0	0
H	0	1	0	0	0	0	0	0

Product Matrix

C_1	1	0	1	1	1	0	0	0
C_2	0	1	0	0	0	1	1	1
H	1	0	0	0	0	0	0	0
H	1	0	0	0	0	0	0	0
H	1	0	0	0	0	0	0	0
H	0	1	0	0	0	0	0	0
H	0	1	0	0	0	0	0	0
H	0	1	0	0	0	0	0	0

FIGURE 3.2 Chemical reaction as matrix addition operation: an ethane initiation example.

Graph of reactant molecule   ⬭ Input

Determine species type

Determine reaction sites
and apply reaction matrices

Determine connected components   Structure/reactivity
                                                   parameters

Determine products uniqueness (isomorphism)
if unique, add to unreacted species list   → Determine reactivity index

Add reaction expression to the network   ← Rate constants expression

Obtain species from unreacted list

FIGURE 3.3 Algorithm for automated generation of reaction network with associated rate constant expressions.

algorithm used to build a reaction network on the computer. The reactions are carried out in a logical manner to make sure all the species only undergo the allowable reaction by checking the specific reaction sites on each species. Every product species formed is then checked for its uniqueness via an isomorphism algorithm to make sure it can be reacted again. When there is no unreacted species left, the network building is complete.

The reaction network is not a reaction model without quantitative rate constants associated. A first-pass estimate of these values is provided by linear free energy relationships (LFERs) (Mochida and Yoneda, 1967 a, b, and c), such as the classic Evans–Polanyi (Evans and Polanyi, 1938) relationship. These are used for correlating rate constants with an appropriate reactivity index of the reacting species for each reaction family. Those reactivity indices can be calculated directly, such as carbon or ring numbers, or indirectly using some computational chemistry package (e.g., MOPAC), such as heat of formation or reaction, from the explicit atom-connectivities of the participating species. The on-the-fly specification of rate constant expressions is also illustrated in Figure 3.3.

These underlying basic algorithms were implemented by Broadbelt et al. (1994a,b, 1996) for developing an ethane pyrolysis model. Significant challenges remained to extend this approach to modeling complex process chemistries with complex, high-carbon-number feedstocks. A major problem is that the reaction network can easily grow beyond the user constraints, even when using cutting-edge computer (CPU, memory, and compiler) capabilities. The problems are enhanced when the algorithms are applied to high-carbon-number reactant compounds or complex multicomponent mixtures for a variety of complex process chemistries. For example, even for a C16 paraffin, there are more than one million isomers; it is not practical or meaningful to keep track of all these isomers.

The biggest challenge is thus how to build a reaction network that can capture the essential chemically and kinetically important species and reactions while keeping the network to a modest size to fulfill the modeling need. There are a variety of ways to handle this conflict. The simple implementation of atom count and rank-based termination criteria used by Broadbelt et al. (1994b) are the first steps to resolving this issue. A system of methodologies has been further exploited for every stage of model building process: preprocessing including rule-based model building and seeding–deseeding ideas, *in situ* processing including on-the-fly generalized isomorphism-based lumping and stochastic sampling, and postprocessing including species-based or reaction-based model reduction and isomorphism-based late lumping. All these model building and control methodologies will be discussed in detail in Section 3.2.

Many candidate reaction networks can be easily built with algorithms on the computer by applying different model building and control methodologies. A fundamental question concerns how similar or different the two reaction networks are, that is, how to characterize a reaction network. A system of algorithms has been developed to characterize species and reactions in the reaction network. More significantly, the generalized isomorphism-based model reduction strategy has been proposed to resolve this issue elegantly by comparing two reaction mechanisms from a lumping point of view. All these network-related properties will be discussed in Section 3.3.

3.2 REACTION NETWORK BUILDING AND CONTROL TECHNIQUES

Every stage of the model building process has been exploited, and a system of methodologies has been developed to show how to build the reaction network that can capture the essential chemically and kinetically important species and reactions while keeping the network to a modest size. For the preprocessing stage, both the rule-based model building strategy and the seeding–deseeding strategy are exploited and developed. The on-the-fly generalized isomorphism-based lumping algorithm and the stochastic sampling algorithm are used to build up the reaction network (i.e., the *in situ* processing stage). After the initial reaction network is formulated (i.e., the postprocessing stage), the species-based or reaction-based model reduction algorithms and the isomorphism-based late lumping strategy reduce the network size while maintaining the chemically and kinetically important species and reactions.

3.2.1 Preprocessing Methodologies

3.2.1.1 Rule-Based Model Building

Reaction rules are the approximations that prune the infinitely large model into its finite, kinetically significant subset. The rules are the differences between vastly different process chemistries that share the same underlying fundamental chemistry (e.g., ethane pyrolysis vs. delayed coking). More significantly, perhaps, they are the key differences between a generic model and its truly useful, often proprietary version. Table 3.1 to Table 3.3 illustrate various generic rules used in the metal-catalyzed, acid-catalyzed, and thermal chemistries, respectively, which

---

TABLE 3.1
Reaction Rules for Metal Chemistry

Metal Reaction Family	Reaction Rule
Dehydrogenation	All positions or limited random selection of sites for saturated compounds (paraffins and naphthenes)
	At the branch and β to branch for isoparaffins and branched naphthenic compounds
Hydrogenation	Hydrogenate at all sites for olefins and olefinic-naphthenes
Saturation	Saturate all aromatic rings to naphthenes in a ring-by-ring manner only
Aromatization	Allow all six-membered naphthenic rings to aromatize
Hydrogenolysis	Formation of light gases (C_1 and C_2) only

TABLE 3.2
Reaction Rules for Acid Chemistry

Acid Reaction Family	Reaction Rule
Protolytic cleavage and hydride abstraction	Deterministic (limited number) or random (treating all sites equally) for saturated compounds (paraffins or naphthenes) At the branch, β to branch, allylic, and β to allylic for isoparaffins and branched naphthenic compounds
Hydride shift, methyl shift, isomerization, ring expansion, and ring contraction	Limited number of reactions as a function of carbon number No formation of geminal branches via isomerization reactions for low acidity processes Isomerization always leads to either increase in branching or increase in the side chain length Formation of more stable ion from a less stable ion ($1° \rightarrow 2° \rightarrow 3°$) only
β-Scission	Allow only A ($3°$ to $3°$) and B ($2°$ to $3°$) type cracking reactions for low acid processes and higher carbon numbers molecules Formation of vinylic (C=C=C) compounds not allowed
Ring closure, ring opening,	Ring closure to either five- or six-membered naphthenic rings Ring opening to stable ions only
Addition	Allowed only for $C_1 \le C_N \le C_4$

TABLE 3.3
Reaction Rules for Thermal Chemistry

Thermal Reaction Family	Reaction Rule
Initiation and hydride abstraction	Deterministic (limited number) or random (treating all sites equally) for saturated compounds (paraffins or naphthenes) At the branch, β to branch, allylic, and β to allylic for iso-paraffins and branched naphthenic compounds
Radical isomerization, ring expansion, and ring contraction	Limited number of reactions as a function of carbon number No formation of geminal branches via isomerization reactions for low temperature processes Isomerization always leads to either increase in branching or increase in the side chain length Formation of more stable radicals from a less stable radical ($1° \rightarrow 2° \rightarrow 3°$) only
β-Scission	Allow β-scissions to allylic, branched or β to branch radicals Formation of vinylic compounds not allowed
Ring closure, ring opening	Ring closure to either five- or six-membered naphthenic rings Ring opening to stable radicals only
Addition	Allowed only for $C_1 \le C_N \le C_4$

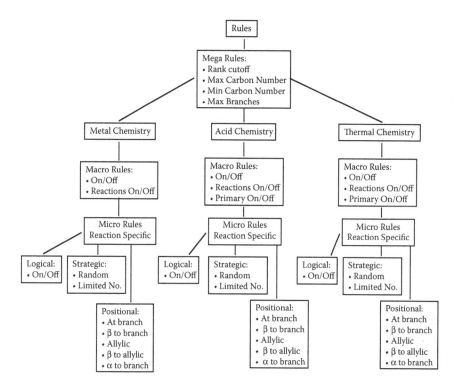

FIGURE 3.4 Classification of rules for construction of complex process chemistry models.

are the most frequently encountered fundamental chemistries occurring in most of the hydrocarbon conversion processes.

A comparison of the rules in Table 3.1 to Table 3.3 exposes the similarities in the rules for the three fundamental chemistries. Figure 3.4 depicts the classification of the rules. The first level of rules, called *mega* rules, specify the rank cutoff and the global carbon number cutoff for all chemistries. The rules for each chemistry are then either chemistry specific (*macro* rules) or reaction specific (*micro* rules). Thus, for a metal chemistry, a macro rule is used to specify whether the chemistry is to be included in the model and what type of reactions are to be considered in the chemistry. Similarly, for the acid and thermal chemistries, apart from specifying the information about inclusion (or exclusion) of various chemistries and reaction families, a macro rule is also used to specify the information about the intermediates considered in the model. For example, primary ions or radicals might be ignored in a model incorporating both acid and thermal chemistries because of their thermochemical instabilities. Micro rules, however, are specific to each reaction in the chemistry and are further classified into three types: *logical*, *strategic*, and *positional*. A logical rule is used to toggle a reaction on or off in a chemistry. A strategic rule gives the information about the approach to be followed, deterministic or random, and whether to have a limit for the total number of reactions, either as a function of rank or carbon number. A positional rule specifies the favored positions for the reaction, either for the reactant or for the product species such as the reaction at the branch, α or β to branch, allylic position, etc.

All the rules are implemented in the model building algorithm to more uniquely define a process. The code and the rules have been separated as much as possible in the algorithm: all the generalized rules are made as user inputs in a user-supplied file instead of hard-coded in the algorithm itself in order to offer users the flexibility to choose. The organization of the knowledge in the form of rules thus enables the users to customize the model building rules with their insights and experience and easily construct customized process models. By choosing a combination of rules, one can easily build a series of models for all kinds of "what-if" scenarios, providing maximum flexibility to test and identify the best reaction network.

3.2.1.2 Seeding and Deseeding

Seeding is a strategy used to direct or guide the model building process toward the empirically observed important products by supplying the network generator with the key intermediates. The essential idea is to guide the model growth to be only in the region of important species and their chemistries. An analogy would be to focus on the trunk and major branches of a tree instead of considering each small branch and leaf.

A contrived situation is helpful to illustrate this idea. For a chemistry that each species can generate n other species through different reactions, there will be a total of $\sum_{r=0}^{r=R} n^r$ species in the reaction network after reacting up to rank R. Clearly, even for a small n (e.g., 4), as R increases (e.g., 10), the model size can grow exponentially and easily go beyond the computer memory and compiler constraints. Seeding involves seeding the model builder with both the reactant and s intermediate species and only allowing the reactions grow to a small rank, R, for reactants to establish the connection to the products via the intermediate species. In this way, the model building will be directed along the seed species pathway and only needs to generate a total of $(1 + s) \sum_{r=0}^{r=\Delta R} n^r$ species in the network. Essentially, the computer builds $1 + s$ smaller low-rank models that, when combined, create a much smaller model than the single one produced by starting with the feed reactants only.

An n-heptane cracking (Watson et al., 1996) can be used to demonstrate the seeding strategy. Figure 3.5 summarizes the chemistry: the primary observable species are paraffins, olefins, and aromatics. The desired aromatics such as benzene, toluene, and xylene are formed at fairly high rank, around 8, from n-heptane. Without seeding, the total number of species that would have been built for R = 8 using n-heptane as a single feed is around 5000, as shown in Figure 3.6. Table 3.4 lists the seeded intermediates along the pathway toward aromatics formation. As we can see from Figure 3.6, the seeded model with R = 3 has 90% fewer species, which will also lead to a much faster solution. Joshi et al. (1998) showed that

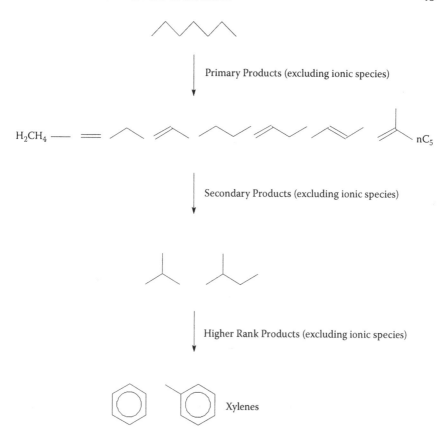

FIGURE 3.5 Reaction network for n-heptane in terms of observable species. (Joshi et al., 1999)

the full and seeded model produced almost identical product spectra and yields, proving that the seeded model captures the essential chemistry and kinetics of the reaction network.

Deseeding, however, facilitates the network building process by supplying the reaction network builder with only some of the reactants in the complex feedstock, rather than the traditional approach of supplying all the reactants in the feedstock to build the reaction network. The reasoning behind this strategy is that after a complex reaction network is built, there will be no absolute meaning of species rank because all the species are interconnected in the network anyway. This is totally different from the reaction network building strategy for a model compound experiment, where the species rank can be defined absolutely relative to the feed reactant. In the complex reaction network derived from a complex feedstock, each node can be treated as a reactant, that is, be assigned with initial concentrations, from the model solving point of view. For example, when we build a hydrocracking model for a wax feed — a highly paraffinic mixture with

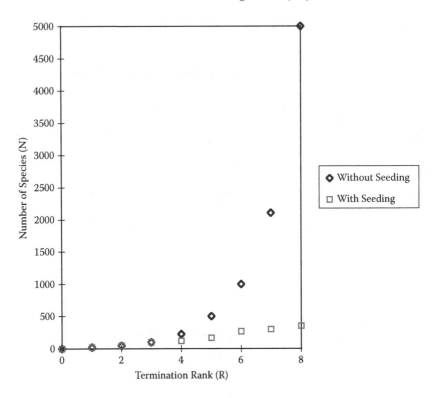

FIGURE 3.6 Comparison of the reaction network with respect to number of species. (Joshi et al., 1999)

a full range carbon number distribution as high as C80 — there is no need to supply the network builder with all the paraffin molecules up to C80. Only one reactant n-C80 molecule would suffice for building up the network because all the other smaller-carbon-number paraffins will be formed through the isomerization and β-scission reactions of higher-carbon-number paraffins anyway, as we will discuss in detail in Chapter 8. A more generalized conclusion from this observation is that, for the purpose of network building, there is no real need to supply all the reactant molecules in the complex feedstock if some of them can be generated from others in the final reaction network. After the whole network is built, there will be no difference between the one using deseeding and the one supplying all the reactant molecules.

Deseeding is not just an easy way to get the work done but is crucial for building a really complex reaction network when the computer memory requirement is being pushed to the limit of the computer hardware capability. This strategy has helped build a C80 hydrocracking model, discussed in Chapter 8, which would not have been possible otherwise.

Seeding and deseeding are very useful for building complex reaction networks. Seeding directs the model building process to capture the essential chemistry with

TABLE 3.4
Seeded Species in the Reaction Network Building of N-Heptane Cracking Toward the Aromatics

Rank	Molecule
2	
4	
6	
8	

observable product species. Deseeding is a new way of looking at the reaction network building process that supersedes the traditional reaction rank understandings and thus enables the model building needed for the complex process chemistries of real complex feedstocks. (Joshi et al., 1999)

3.2.2 *IN SITU* PROCESSING METHODOLOGIES

3.2.2.1 Generalized Isomorphism Algorithm as an On-the-Fly Lumping Tool

It is useful to analyze why the automated reaction network generation is so easy to explode for complex process chemistry with complex feedstocks. Recall from the automated model building algorithm in Figure 3.3 that every time a product species is generated, its isomorphism is checked for its uniqueness, and then it is added to the unreacted species list for further reaction. This isomorphism algorithm (Broadbelt et al., 1996) identifies each species at the atomic connectivity level using a graph-theoretic approach. For example, if only one n-hexadecane reactant is fed in a hydrocracking process, it will, in principle, generate more than one million possible C16 isomers at the molecular level on the computer (even without explicit accounting of all intermediate carbenium ions). This can easily go beyond the memory and solving capabilities of state-of-the-art computers.

A fundamental question arising from this challenge is whether we really should or need to take into account all the possible isomers in the detailed molecular modeling approach. For a catalyst surface study using a small model compound, the answer is probably yes. In this case, we would like to keep track of all the possible species in the experiment (e.g., no more than hundreds) and analyze the influence of the properties of the catalyst on the product spectrum. The isomorphism algorithm (Broadbelt et al., 1996) identifying each species at the atomic connectivity level makes sense in this case. However, in the process modeling of complex chemistries with complex feedstocks (e.g., gas oil hydrocracking), it is impossible, and there is really no need even in a detailed kinetic model to keep track of all the possible species in the process stream, which has gone beyond the capability of current available computer and analytical technology. The right balanced modeling strategy lies between the molecule-based fundamental chemistries and the computer and analytical capabilities.

After realizing that the isomorphism algorithm identifies the main reason for automated network building explosion for complex process chemistries, the next question asks when two species should be considered the same or can be grouped together at what level without sacrificing the fundamental mechanism. There are many choices for the criteria: empirical formula, group additivity, energetics, molecular weight or some other analytical property, etc. The essential goal of a detailed molecule-based model is to enable the user to predict the product property or performance from the molecular detail in the model. Therefore, considering the overall modeling strategy discussed in Chapter 1, one possible answer would be that two species could be grouped together at the available quantitative structure/ property relationship (QSPR) level. This is because even if we have the atomic connectivity-level detail of all the species, if the QSPRs used to estimate the reactivity and property are only at a group contribution level, there will be no real benefit from keeping track of all the atomic details of all species. For example, in the hydrocracking process, all the polynuclear aromatics are hydrogenated in a ring-by-ring manner, and the corresponding QSPR for the ring saturation reaction is only correlated at an aromatic ring number and saturated carbon number level (Korre, 1995). Then the molecules with same aromatic rings and saturated carbons may be lumped together. One representative molecule for this structure may be sufficient for the modeling purpose. As another example for heavy paraffin hydrocracking (discussed in Chapter 8), the carbon number and branch number level may be enough for our modeling objective because there is so far no easy way to distinguish, for example, the tri-branched C60 isomers using currently available analytical techniques.

Based on these understandings, a generalized isomorphism algorithm was developed to accommodate the isomorphism check at different levels. The user can choose or supply criteria for different species and reaction types, ranging from empirical formulae to atomic connectivity level. For example, for paraffins, the combination of carbon number and branch number can be an important criteria; for ring compounds, the combination of core ring structure and side-chain number and length is important; for sulfur compounds, key positions like α to the sulfur on thiophenic compounds are important. The criteria can be different for different process chemistries and can be customized by the user based on his or her insights and experience. The same reaction that forms via the generalized isomorphism algorithm is added up to the reaction network to ensure that the reaction pathway degeneracy (RPD) is maintained.

This generalized isomorphism-based lumping strategy has fundamentally superseded classical lumping schemes. There is no fundamental chemistry and molecular composition in the traditional lumped models. However, the automated network generation builder embedded with the generalized isomorphism algorithm enables the user to derive molecule-based models as needed from the fundamental chemistries and reaction mechanisms and at the same time provides the ultimate flexibility to lump and reduces the model size on-the-fly.

This model building strategy can thus be used to generate a whole spectrum of custom molecule-based models derived from the fundamental chemistries. This isomorphism-based lumping strategy is crucial in many cases. For example, the kinetic model is normally only the reaction part of a more complex computational fluid dynamics (CFD) model. In this case, a small (e.g., 20 to 30 lumps), but fundamentally solid kinetic scheme is needed because of the computational complexity and CPU requirement. This can be easily achieved with our generalized isomorphism algorithm by defining the isomorphic criteria and deriving a concise but fundamentally solid kinetic model. The *in situ* lumping flexibility is also crucial for making the real complex model building possible, as we can see from the heavy paraffin hydrocracking model in Chapter 8, because otherwise there is no way to build a model with all possible isomers on the computer.

3.2.2.2 Stochastic Rules for Reaction Site Sampling

The basic idea of stochastic model building is to sample reactions stochastically on each reaction family and each compound type. This works especially for a wide range of molecular mixtures because of the way the molecules compensate for each other in the mixture. For example, we consider hydrocracking of a wax feedstock with a wide range distribution of paraffins of each carbon number. We can only sample the reaction sites on a long-chain paraffin (e.g., C60), so it can be isomerized and cracked to form smaller paraffins, which may not the full spectrum. However, if we apply the sampling strategy to a wide range of long-chain paraffins, a full spectrum of products can still be formed because of the compensation effects between them.

The probability density function (PDF) discussed in Chapter 2 can also be used to describe the reaction probabilities at various sites. We can assume the distribution function to sample or select the possible reactions for each reaction family and compound class. Then, by matching the product spectrum with the experimental observations, we can optimize the parameters for each PDF. This methodology can also help us build a reaction network with the essential chemistries and reaction mechanisms embedded in it.

The stochastically formed reaction network is just part of the theoretically complete reaction network. One way to recover the rate information from this small representative network to the full one is by maintaining the RPD for each reaction family on each species. This is equivalent to the normalization of rate constants, so the sampled reactions can be used to represent the essential kinetic effects of the whole spectrum of reactions on that species.

Mizan et al. (1998) has implemented this methodology to successfully build the hydroisomerization reaction networks for various waxes.

3.2.3 POSTPROCESSING METHODOLOGIES

3.2.3.1 Generalized Isomorphism-Based Late Lumping

The generalized isomorphism-based lumping discussed in Section 3.2.2.1 can be used even after a reaction network has been built. As a reaction network is being built, all the information about each species, such as its carbon number, Z number, structural groups, molecular weight, boiling point, species type, etc., is obtained and recorded into the database files. After a complete reaction network has been built, we can always group the species and reactions together and form a smaller model. By specifying different isomorphism criteria, users can customize the complete model to any level of lumped model matching the analytical capabilities or data availability. For example, if the available experimental data for paraffin hydrocracking is only at carbon number and branch number level, it will be practical to lump the model to this level and easy to tune this smaller model first. The lumped model can be solved fast, so it can be tuned with less effort. These isomorphism-based lumped models are derived from the fundamental chemistries and reaction mechanisms and can still be on a molecular basis as needed, which is fundamentally different from the randomly selected lumping schema.

This generalized isomorphism-based late lumping strategy is flexible enough to customize the most detailed fundamental model and deliver the right size and informative model as needed. This also suggests an adaptive model tuning strategy — any developed model can always be tuned with the available data and improved along the way as more data become available. This is significant because the current model development paradigm has completely changed because automated model building is so fast and efficient compared with model tuning or parameter optimization. An adaptive model tuning strategy can speed up the whole model development process.

3.2.3.2 Species-Based and Reaction-Based Model Reduction

We have also exploited model reduction methodologies based on sensitivity analysis even after the model has been tuned, that is, after the rate parameters have been optimized. To understand these methods, it is important to differentiate the type of species in the model: *important, necessary,* and *redundant.* Important species are of direct importance to the user. Necessary species are those the model must retain in order to correctly model the important species. Redundant species are those that can be eliminated from the model without a large effect to the prediction of the important species. The user must define the set of important species and an acceptable level of error in these species; the model reduction algorithms then choose the necessary and redundant species.

Turanyi (1990) proposed a method of model reduction that is an iterative scheme involving sensitivity analysis of rate of production of important species with respect to concentrations of all other species. He defined the vector B to have elements

$$B_i = \sum_{n=1}^{N} \left(\frac{\partial \ln r_n}{\partial \ln c_i} \right)^2 \qquad (3.1)$$

where the sum was carried out over the set of important and necessary species. Species possessing the highest B_i values were included in the set of necessary species, and the process was repeated until the vector B converged. Redundant species are those not considered necessary or important after convergence occurs.

A generalized species-based algorithm for model reduction building upon the above method has been developed and is described as follows:

1. Start with an empty set of necessary species. Select the set of important species, a set of times of interest, and a fractional cutoff (between 0 and 1) to be used later. Solve the full model for the concentrations of each species at each time of interest.
2. Calculate the sensitivity of the rates of production of important and necessary species to changes in the concentration of all other species in the reaction network. Call this measure of error Φ_j for species j.
3. Determine the range of Φ_j and compute a cutoff value Φ_c that is the fractional cutoff (as selected in step 1) of the Φ_j range. Add all species with $\Phi_j > \Phi_c$ to the necessary species set.
4. Repeat steps 2 and 3 until no new necessary species is discovered.
5. Construct the reduced model using all reactions that consume the important and necessary species.

Similarly, a reaction-based model reduction algorithm is developed as follows:

1. Start with an empty set of necessary species. Select the set of important species, a set of times of interest, and a fractional cutoff (between 0 and 1) to be used later. Solve the full model for the concentrations of each species at each time of interest.
2. Calculate the sensitivity of the rates of production of important and necessary species to changes in the rate parameters of each reaction. Call this measure of error Φ_j for reaction j.

3. Determine the range of Φ_j and compute a cutoff value Φ_c that is the fractional cutoff (as selected in step 1) of the Φ_j range. Add all species that are consumed by reactions with $\Phi_j > \Phi_c$ to the necessary species set.
4. Repeat steps 2 and 3 until no new necessary species is discovered.
5. Construct the reduced model using all reactions that have a Φ_j value greater than a given cutoff.

Different species-based or reaction-based model reduction techniques can be applied by defining an appropriate function to determine the sensitivity of the important and necessary species to each of the other species or reactions as required in step 2 of both algorithms. Two such functions are sensitivity analysis and the "on/off" method. Using sensitivity analysis, the function for quantifying the effect one reaction has on the set of important and necessary species can be

$$\Phi_j = \sum_{}^{times} \sum_i^{species} \left(\frac{\partial r_i}{\partial k_j} \right)^2 \tag{3.2}$$

where the species in the inner sum are the members of important and necessary species. Thus, Φ_j carries a measure of the instantaneous sensitivity of the rate of production of the important and necessary species on the kinetic parameter of reaction j at a specified set of times. The on/off method sets the kinetic parameter or concentration being analyzed to zero and examines the effect this change has on the predictions made by the model. The function to be analyzed is

$$\Phi_j = \sum_{}^{times} \sum_i^{species} (r_i^* - r_i)^2 \tag{3.3}$$

where a superscript * represents the reduced model, and the sum over the species is carried over only the important and necessary ones. Both the species-based and reaction-based cases use the concentration profiles of the full model as a basis when computing r_i^*, so the reduced model does not have to be solved.

Families of reduced models can thus be created using this methodology by changing only the fractional cutoff. Generally, since the implications of a particular choice of the cutoff fraction are not apparent *a priori*, a range of cutoff fractions should be explored to locate the optimal one. A plateau in overall error is a general feature of a graph of overall error vs. number of species or number of reactions in the reduced model, and the smallest model on this plateau marks an optimum reduced model if no target accuracy is available.

Baynes (1997) applied these algorithms on an n-hexane acid-cracking mechanistic model with 597 species and 2932 reactions and reduced the model to 390 species and 1342 reactions. This correctly models the concentrations of the specified important species and only takes 33% of the time required to solve the full one. An important finding from these tests is that the reaction-based model reduction performs better than the species-based one. If an important species undergoes both fast and slow reactions, reaction-based reduction can remove the slower reactions, but species-based reduction cannot. The trade-off is that the reaction-based model reduction takes longer to complete because it analyzes more possibilities.

The above algorithms can also be modified and used as *in situ* model reduction tools to identify redundant species or reactions and control the mechanism growth in the automated generation of networks if the rate parameters are already known and the reaction rate can be evaluated on-the-fly. In the thermal cracking process, there is a relatively good idea of the rate parameters, and this can be incorporated and used to control the model growth adaptively. However, in most of the heterogeneous catalytic processes, the rate parameters are normally not known *a priori*, so the developed algorithms can only be used as postprocessing tools after a model is tuned.

3.3 PROPERTIES OF REACTION NETWORKS

3.3.1 PROPERTIES OF SPECIES

Each species is a node in the reaction network. It is very important to generate all the properties of each species during the reaction network building. In the automated network building algorithm, each species is built up from the atom level, and its atomic connectivity is clear. Therefore, most of the property of the species can be identified or calculated. Various atoms including C, H, S, N, and O can be counted. Thus, the Z number (defined as 2* carbon number − hydrogen number) can be easily calculated as an important measure of the unsaturation of a compound. The molecular weight can also be calculated directly from atom counts.

The species type can be determined through its structure. Figure 3.7 shows the classification of species type that has been implemented in the algorithm. For different types of species, different classes of reactions can be carried out accordingly.

Many properties of species can be calculated from their structures through the molecular structure–property correlations or computational chemistry packages. For example, the boiling point can be predicted through the group additivity method or correlated with the structures for each compound class. Similarly, the density of each species can be predicted through many correlations with molecular structures. With boiling point and density information calculated from the molecular structures, many global properties of the petroleum mixture can be estimated accurately from its compositions (Quann and Jaffe, 1996).

In addition, a unique user-friendly name for each species is generated along the reaction network building. This is implemented through a hierarchical approach. If there is an IUPAC (International Union of Pure and Applied

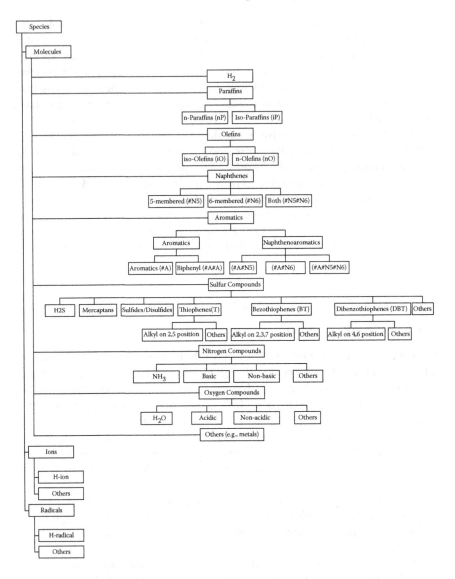

FIGURE 3.7 Classification of species type (# denotes ring number).

Chemisty) name for the generated species in our developed database that has all the important species, it will be used. Otherwise, the molecular structure of the species is analyzed and described using the structural string with a sorted ID unique to the database. The user of the model can have a clear idea of the species type and the detailed structure of each species, which is normally sufficient for modeling.

For each species, all the reactions it participates in can be sorted after the reaction network is built. Therefore, the reaction coordination number for each species can also be easily counted and analyzed.

3.3.2 PROPERTIES OF REACTIONS

Each reaction is an edge in the reaction network. It is very important to organize the reactions into reaction families. Each reaction family can be described mathematically through the succinct reaction matrix and can be easily implemented to carry out the chemical reactions on the computer. Furthermore, the reaction family can be used to organize the rate information, as will be discussed in Chapter 4. Table 3.5 lists the commonly used reaction families that have been implemented in the generic automated network building algorithm at the mechanistic level.

The supporting code used to organize and analyze the reactions is also developed. All the reactions in the network can be sorted for each reaction family, and they can be sorted by each species; all the reactions involving any specified species can be organized for users' convenience.

TABLE 3.5
Reaction Families for Metal, Acid, and Thermal Chemistries

Metal	Acid	Thermal
Molecular Level	**Bimolecular**	**Bimolecular**
Hydrogenation	Protonation	Bond fission
Dehydrogenation	Protolytic cleavage	H-abstraction
Aromatization	Hydride abstraction	Recombination
Saturation	Addition	Termination
Hydrogenolysis		Addition
	Unimolecular	**Unimolecular**
	Deprotonation	β-Scission
	β-Scission	Ring closure
	Isomerization	Ring opening
	Hydride shift	Radical isomerization
	Methyl shift	
	Ring expansion	
	Ring contraction	
	Ring closure	
	Ring opening	

3.3.3 CHARACTERIZATION OF THE REACTION NETWORK

Just like a species is a graph, with atoms being the nodes and bonds being the edges, the reaction network is a graph, with species being the nodes and reactions being the edges. Therefore, after we have defined a unique name or ID for each species in the network, all the generic algorithms that have been developed for the molecular graph can also be applied to the reaction network graph.

However, the detailed reaction network for the process chemistry normally has hundreds of species and thousands of reactions. It would be computationally infeasible to apply the same algorithms directly to the network graph. Moreover, the bimolecular reactions in the reaction network make the graph-theoretical algorithms much more difficult. Temkin et al. (1996) have analyzed the chemical reaction networks from the mathematical topology and graph theory point of view and proposed useful mathematical indices to characterize the reaction networks.

The generalized isomorphism algorithm and understanding of the reaction networks partly characterize the property of a reaction network from the lumped point of view. Although it is difficult to tell quantitatively how different two reaction networks are, both can easily be lumped to smaller networks by applying the isomorphism-based lumping strategy and thus can be easily compared. Depending on the isomorphic criteria, we can easily compare two reaction networks and determine if they are the same or similar at some specified level. For example, two stochastically built paraffin hydrocracking reaction networks are totally different from each other on the species basis, but they could be same at the carbon number and branch number levels. If what we are concerned with is the carbon number distribution and iso/normal ratio predictions from the model, the two reaction networks can be treated as equivalent.

3.4 SUMMARY AND CONCLUSIONS

The graph-theoretical concepts that enable automated reaction model construction are reviewed. A molecule can be represented by a chemical graph, with the atoms being the nodes and the bonds being the edges. The bond–electron matrix representation enables a chemical reaction to be implemented through a simple matrix addition operation. Through repetitive application of reaction matrices to the reactants and their progeny, the reaction network can then be built. By applying the quantitative rate constants associated with it, the reaction network becomes a reaction model.

However, a major problem with the automated reaction network building is that when the same algorithms are applied to high-carbon-number reactant compounds or complex multicomponent mixtures for a variety of complex process chemistries, the network can easily explode and grow beyond the user constraints and computer (CPU, memory, and compiler) capabilities. The biggest challenge is how to build the reaction network that can capture the essential chemically and kinetically important species and reactions while keeping the network to a modest size.

Every stage of the model building process has been exploited, and a system of methodologies has been developed to guide the building of the reaction network.

In the preprocessing stage, both the rule-based model building strategy and the seeding/deseeding strategy are exploited and developed to provide users with the flexibility to guide the model building process with their expertise. In the *in situ* processing stage, we have extended the original isomorphism algorithm and created the on-the-fly generalized isomorphism-based lumping algorithm and the stochastic sampling algorithm. In the postprocessing stage, we have also developed the species-based or reaction-based model reduction algorithms and the isomorphism-based late lumping strategy to reduce the network size.

The generalized isomorphism-based lumping strategy provides users with the maximum flexibility of building molecule-based reaction models on the basis of fundamental chemistry and reaction mechanisms by balancing the needs with the available data and QSPRs. This also helps answer the fundamental question of how similar or different the two reaction networks are by comparing two reaction mechanisms from an isomorphic-lumping point of view. All the network-related properties are also characterized and discussed.

REFERENCES

Baynes, B., BS thesis, University of Delaware, Newark, 1997.

Broadbelt, L.J., Stark, S.M., and Klein, M.T., Computer generated reaction networks: on-the-fly calculation of species properties using computational quantum chemistry, *Chem. Eng. Sci.*, 49, 4991–5101, 1994a.

Broadbelt, L.J., Stark, S.M., and Klein, M.T., Computer generated pyrolysis modeling: on-the-fly generation of species, reactions and rates, *Ind. Eng. Chem. Res.*, 33, 790–799, 1994b.

Broadbelt, L.J., Stark, S.M., and Klein, M.T., Computer generated reaction modeling: decomposition and encoding algorithms for determining species uniqueness, *Comput. Chem. Eng.*, 20, 113–129, 1996.

Evans, M.G. and Polyani, M., Inertia and driving force of chemical reactions, *Trans. Faraday Soc.*, 34, 11, 1938.

Joshi, P.V., Molecular and Mechanistic Modeling of Complex Process Chemistries, Ph.D. dissertation, University of Delaware, Newark, 1998.

Joshi, P.V., Freund, H., Klein, M.T., Directed Kinetic Model Building: Seeding as a model reduction tool, *Energy Fuel*, 13(4), 877–880, 1999.

Korre, S.C., Quantitative Structure/Reactivity Correlations as a Reaction Engineering Tool: Applications to Hydrocracking of Polynuclear Aromatics, Ph.D. thesis, University of Delaware, 1995.

Mizan, T.I., Hou, G., and Klein, M.T., Mechanistic Modeling of the Hydroisomerization of High Carbon Number Waxes, AIChE Meeting New Orleans, paper 25a, 1998.

Mochida, I. and Yoneda, Y., Linear free energy relationships in heterogeneous catalysis. I. Dealkylation of alkylbenzenes on cracking catalysts, *J. Catal.*, 7, 386–392, 1967a.

Mochida, I. and Yoneda, Y., Linear free energy relationships in heterogeneous catalysis. II. Dealkylation and isomerization reactions on various solid acid catalysts, *J. Catal.*, 7, 393–396, 1967b.

Mochida, I. and Yoneda, Y. Linear free energy relationships in heterogeneous catalysis. III. Temperature effects in dealkylation of alkylbenzenes on the cracking catalysts, *J. Catal.*, 7, 223–230, 1967c.

Quann, R.J. and Jaffe, S.B., Structure-oriented lumping: describing the chemistry of complex hydrocarbon mixtures, *Ind. Eng. Chem. Res.*, 31(11), 2483, 1992.

Quann, R.J. and Jaffe, S.B., Building useful models of complex reaction systems in petroleum refining, *Chem. Eng. Sci.*, 51(10), 1615, 1996.

Temkin, O.N., Zeigarnik, A.V., and Bonchev, D., *Chemical Reaction Networks: A Graph-Theoretical Approach*, CRC Press, Boca Raton, FL, 1996.

Turanyi, T., Reduction of large reaction mechanisms, *New J. Chem.*, 14, 795–803, 1990.

Ugi, I., Bauer, J., Brandt, J., Freidrich, J., Gasteiger, J., Jochum, C., and Schubert, W. New applications of computers in chemistry, *Agnew. Chem. Int. Ed. Engl.*, 18, 111–123, 1979.

Watson, B.A., Klein, M.T., and Harding, R.H., Mechanistic modeling of n-heptane cracking on HZSM-5, *Ind. Eng. Chem. Res.*, 35(5), 1506–1516, 1996.

4 Organizing Kinetic Model Parameters

4.1 INTRODUCTION

The previous chapter covered various techniques for the automated construction of detailed reaction networks of complex process chemistries. The reaction network by itself, however, is not a complete description of the complex process chemistry. The associated rate and equilibrium parameters are required. In this chapter, we will focus on the methodologies of organizing the parameters of detailed kinetic models.

An inherent problem associated with detailed kinetic modeling is that there can be thousands of reactions and therefore thousands of rate constants associated with the reaction network. It is impossible to evaluate the rate constant of each reaction through thousands of model compound experiments or reconcile so many rate constant parameters with the experimental data at the same time using currently available optimization algorithms. It is thus clear that approaches aimed at reducing the complexity and number of model parameters would be of great value in the formulation and use of detailed molecular models of complex mixtures.

Various levels of approaches can handle this problem. A first-order of approximation is that each reaction family is assumed to have the same kinetic parameters since all the reactions undergo similar intramolecular rearrangement. However, this is generally insufficient, considering the perturbation on the reaction rates of a wide range of molecular structures even for the same reaction family. The other extreme is the direct computation of rate constants from first principles through the implementation of quantum chemical calculations. While the quantum mechanics theory has grown to an unprecedented level, the molecular size it can calculate in a feasible amount of time is still limited.

A practical resolution of this conflict emerges from a more careful scrutiny of the composition and reactions of complex feedstocks. Much of the complexity is statistical. Each of the $O(10^5)$ species in the complex mixture falls into one of a handful of compound classes (e.g., paraffins, olefins, naphthenes, aromatics, alkylaromatics), and these in turn react in a manageable set of reaction families (e.g., hydrogenation, isomerization, dealkylation). Thus, the complexity is really the simultaneous reaction of sets of many similar compounds. Within each set, compounds differ only in substituents, and differences in reactivity are attributable to these substituents. This suggests the use of quantitative structure/property relationships (QSPRs) as an organizational and predictive technique for the parameters in detailed kinetic modeling.

In general, the concept that there is a relationship between the properties of compounds and their molecular structures is inherent to chemistry. It is the basic tenet of chemistry to attempt to identify these relationships between molecular structure and property and to quantify them. Hammett (1937) organized reaction families and substituent effects for homogeneous systems. The Evans–Polyani (Evans and Polyani, 1938) relationship ($E^* = E_0^* + \alpha\Delta H$) is another classic example. The basic premise behind the quantitative structure reactivity correlation (QSRC) is that the rate constants for a defined reaction family can be correlated with some molecular property for a homologous series of reactions. The simplest form of these relationships is

$$\ln k_i = a + b * RI_i \tag{4.1}$$

where k_i is the rate constant of molecule i, RI is the reactivity index for molecule i, and a and b are the correlation coefficients to be determined that are specific to some reaction family. This relationship is very well known as a linear free energy relationship (LFER). Mochida and Yoneda (1967a,b, and c) first demonstrated the use of LFERs for catalytic reactions. Klein and coworkers (Neurock, 1992; Korre, 1995) have extensively developed the LFERs to correlate reaction rates of various metal- and acid-catalytic chemistries as well as free-radical chemistry. If a and b in Equation 4.1 can be determined experimentally from several representative reactions, the rate constants for all members in the same reaction family can be calculated from reactivity indices that can in turn be evaluated from the molecular structures. LFERs have also been developed as a convenient way to organize the equilibrium and adsorption constants (Korre, 1995).

In this chapter, the kinetic rate laws of the complex reaction network are briefly discussed at both the pathways level and mechanistic level in Section 4.2. The development and examples of classical LFER concept will be reviewed in Section 4.3, and an extensive summary of developed LFERs for hydrocracking chemistry is presented in Section 4.4. Finally, in Section 4.5, the LFER concept is extended and generalized. The LFER is extended through a Taylor series approach to the general QSRC with molecular structures. A more generalized and practical QSRC methodology is proposed to correlate the rate parameters in the detailed kinetic model development process.

4.2 RATE LAWS FOR COMPLEX REACTION NETWORKS

There are two levels of detailed kinetic modeling: the pathways level and the mechanistic level. At the pathways level, only the observable species in the process are taken into account, and all the intermediates, such as the radicals or ions, are implicit: the reaction network only describes the observable reaction pathways between the observable molecules. At the mechanistic level, in principle, all the elementary reactions in the complex process are counted in the reaction network. For example, in the mechanistic level modeling of thermal cracking, all the reactions are carried out based on the free radical mechanism, where all the molecules and free radicals are explicitly described in the reaction network. Another example is mechanistic level modeling of catalytic cracking, in which all the reactions are carried out based on the carbenium ion chemistry mechanism, where all the molecules and carbenium ions are explicitly described in the reaction network.

4.2.1 KINETIC RATE LAWS AT THE PATHWAYS LEVEL

For homogeneous systems, the rate law formula is simple as long as the reaction order is determined. The best way to determine the reaction order is through experiments. In detailed kinetic modeling, other compounds in a given homologous series will generally be assumed to follow the same reaction order for the same reaction family. For each reaction, the rate law will be written as the simple expression Equation 4.2 for the reaction $A + B \leftrightarrow R + S$:

$$r = k \, (C_A C_B - C_R C_S/K) \tag{4.2}$$

where k is the rate constant; K is the equilibrium constant; α, β, γ, and δ are reaction orders; and C_i is the concentration of compound i.

For the heterogeneously catalyzed systems, the rate law is more complete. To date, the conventional way to describe the catalytic kinetics is using the Langmuir–Hinshelwood–Hougen–Watson (LHHW) formalism. Froment and Bischoff (1990) systematically discussed and summarized the classical LHHW formalism to describe the heterogeneous rate laws. The generic formula is

$$rate = \frac{(\text{Kinetic Group})(\text{Driving Force Group})}{(\text{Adsorption Group})^n} \tag{4.3}$$

where each group for reactions on solid catalysts was developed by Yang and Hougen (1950). For a complete reference of detailed kinetic modeling in this book, we reproduced Table 4.1 from Froment and Bischoff (1990). With permission.

For example, for the bimolecular reaction $A + B \leftrightarrow R + S$, when surface reaction is controlling,

$$r = \frac{k_{sr} K_A K_B [p_A p_B - (p_R p_S/K)]}{(1 + K_A p_A + K_B p_B + K_R p_R + K_S p_S + K_I p_I)^2} \tag{4.4}$$

where k_{sr} is the surface reaction rate, K is the reaction equilibrium constant, K_i is the adsorption constant of component i, p_i is the partial pressure of component i, I is any adsorbable inert, and the exponent 2 indicates that both reactants are adsorbed on the catalyst surface.

TABLE 4.1
LHHW Formalism (Yang and Hougen, 1950)

Kinetic Groups

Adsorption of A controlling	k_A
Adsorption of B controlling	k_B
Desorption of R controlling	$k_R K$
Adsorption of A controlling with dissociation	k_A
Impact of A controlling	$k_A\,K_B$
Homogeneous reaction controlling	k

Surface Reaction Controlling

	$A \leftrightarrow R$	$A \leftrightarrow R + S$	$A + B \leftrightarrow R$	$A + B \leftrightarrow R + S$
Without dissociation	$k_{sr}\,K_A$	$k_{sr}\,K_A$	$k_{sr}\,K_A K_B$	$k_{sr}\,K_A K_B$
Without dissociation of A	$k_{sr}\,K_A$	$k_{sr}\,K_A$	$k_{sr}\,K_A K_B$	$k_{sr}\,K_A K_B$
B not adsorbed	$k_{sr}\,K_A$	$k_{sr}\,K_A$	$k_{sr}\,K_A$	$k_{sr}\,K_A$
B not adsorbed, A dissociated	$k_{sr}\,K_A$	$k_{sr}\,K_A$	$k_{sr}\,K_A$	$k_{sr}\,K_A$

Driving-Force Groups

Reaction	$A \leftrightarrow R$	$A \leftrightarrow R + S$	$A + B \leftrightarrow R$	$A + B \leftrightarrow R + S$
Adsorption of A controlling	$p_A - \dfrac{p_R}{K}$	$p_A - \dfrac{p_R p_S}{K}$	$p_A - \dfrac{p_R}{K p_B}$	$p_A - \dfrac{p_R p_S}{K p_B}$
Adsorption of B controlling	0	0	$p_B - \dfrac{p_R}{K p_A}$	$p_B - \dfrac{p_R p_S}{K p_A}$
Desorption of R controlling	$p_A - \dfrac{p_R}{K}$	$\dfrac{p_A}{p_S} - \dfrac{p_R}{K}$	$p_A p_B - \dfrac{p_R}{K}$	$\dfrac{p_A p_B}{p_S} - \dfrac{p_R}{K}$
Surface reaction controlling	$p_A - \dfrac{p_R}{K}$	$p_A - \dfrac{p_R p_S}{K}$	$p_A p_B - \dfrac{p_R}{K}$	$p_A p_B - \dfrac{p_R p_S}{K}$
Impact of A controlling (A not adsorbed)	0	0	$p_A p_B - \dfrac{p_R}{K}$	$p_A p_B - \dfrac{p_R p_S}{K}$
Homogeneous reaction controlling	$p_A - \dfrac{p_R}{K}$	$p_A - \dfrac{p_R p_S}{K}$	$p_A p_B - \dfrac{p_R}{K}$	$p_A p_B - \dfrac{p_R p_S}{K}$

TABLE 4.1(Continued)
LHHW Formalism (Yang and Hougen, 1950)

Replacements in the General Adsorption Groups
$$(1 + K_A p_A + K_B p_B + K_R p_R + K_S p_S + K_i p_i)^n$$

Reaction	$A \leftrightarrow R$	$A \leftrightarrow R + A$	$A + B \leftrightarrow R$	$A + B \leftrightarrow R + S$
Where adsorption of A is rate controlling, replace $K_A p_A$ by	$\dfrac{K_A p_R}{K}$	$\dfrac{K_A p_R p_S}{K}$	$\dfrac{K_A p_R}{K p_B}$	$\dfrac{K_A p_R p_S}{K p_B}$
Where adsorption of B is rate controlling, replace $K_B p_B$ by	0	0	$\dfrac{K_B p_R}{K p_A}$	$\dfrac{K_B p_R p_S}{K p_A}$
Where desorption of R is rate controlling, replace $K_R p_R$ by	$KK_R p_A$	$KK_R \dfrac{p_A}{p_S}$	$KK_R p_A p_B$	$KK_R \dfrac{p_A p_B}{p_S}$
Where adsorption of A is rate controlling with dissociation of A, replace $K_A p_A$ by	$\sqrt{\dfrac{K_A p_R}{K}}$	$\sqrt{\dfrac{K_A p_R p_S}{K}}$	$\sqrt{\dfrac{K_A p_R}{K p_S}}$	$\sqrt{\dfrac{K_A p_R p_S}{K p_A}}$
Where equilibrium adsorption of A takes place with dissociation of A, replace $K_A p_A$ by (and similarly for other components adsorbed with dissociation)	$\sqrt{K_A p_A}$	$\sqrt{K_A p_A}$	$\sqrt{K_A p_A}$	$\sqrt{K_A p_A}$
Where A is not adsorbed, replace $K_A p_A$ by (and similarly for other components that are not adsorbed)	0	0	0	0

Exponents of Adsorption Groups

Adsorption of A controlling without dissociation	$n = 1$
Desorption of R controlling	$n = 1$
Adsorption of A controlling with dissociation	$n = 2$
Impact of A without dissociation $A + B <-> R$	$n = 1$
Impact of A without dissociation $A + B <-> R + S$	$n = 2$
Homogeneous reaction	$n = 0$

Surface Reaction Controlling

Reaction	$A \leftrightarrow R$	$A \leftrightarrow R + S$	$A + B \leftrightarrow R$	$A + B \leftrightarrow R + S$
No dissociation of A	1	2	2	2
Dissociation of A	2	2	3	3
Dissociation of A (B not adsorbed)	2	2	2	2
No dissociation of A (B not adsorbed)	1	2	1	2

To extend the LHHW formalism to complex process chemistry involving hundreds or thousands of components for any reaction in the complex reaction network, the denominator adsorption group should extend to $(1 + \sum K_i p_i)$ to take into account the total inhibition effect of all the components on the reactive catalytic site. For the catalytic systems containing different active sites for different reactions, the rate law of Equation 4.3 should be formulated separately for different sites. For example, Korre (1995) successfully developed the following dual site LHHW rate law for the polynuclear aromatics (PNA) hydrocracking chemistry to account for the dual function (both metal and acid) of the hydrocracking catalyst. The rate for each product was obtained as a summation of the rates of its transformations on the metal and the acid sites:

$$\frac{dC_i}{dt} = \frac{\sum_j k_{ji}(C_j - C_i / K_{ji})}{D_H} + \frac{\sum_l k_{li}(C_l - C_i / K_{li})}{D_A} \tag{4.5}$$

where, C_i, C_j, and C_l are component concentrations (mol l^{-1}); k_{ji} and k_{li} are combined numerator rate parameters (1 kg$_{cat}^{-1}$ s^{-1}) (including intrinsic rate and adsorption constant contributions and hydrogen pressure where applicable), and K_{ji} and K_{li} are the equilibrium ratios (mol$_i$/mol$_j$/P$_{H2n}$). D_H and D_A are the adsorption groups for the metal (hydrogenation, D_H) and zeolite (acid, D_A) sites, respectively, defined in Equation 4.6:

$$D_H = 1 + \sum_i K_i^H C_i, \quad D_A = 1 + \sum_i K_i^A C_i \tag{4.6}$$

where C_i represents component concentration (mol l^{-1}), and K_i^H and K_i^A (1 mol^{-1}) represent individual component adsorption constants on the metal and acid sites, respectively. Implicit assumptions are surface reaction as the rate determining step and unity adsorption exponent group in all cases.

Another example, shown in Chapter 9, is the "two site" (σ site for hydrogenolysis and π site for hydrogenation) hydrodesulfurization (HDS) kinetic rate law proposed as follows for every thiophenic compound (thiophene, benzothiophene, dibenzothiophene, and their alkyl derivatives):

$$r = \frac{f_\sigma k K_{A,\sigma} K_{H,\sigma}[A][H_2]}{(1 + \sum_i K_{i,\sigma}[I] + \sqrt{K_{H,\sigma}[H_2]}\,)^n} + \frac{f_\sigma k_\sigma K_{A,\sigma} K_{H,\sigma}([A][H_2] - [B]/K)}{(1 + \sum_i K_{i,\sigma}[I] + \sqrt{K_{H,\sigma}[H_2]}\,)^n} \tag{4.7}$$

where [I] is the concentration of component I, f is the global steric and electronic factors for alkyl substituents with respect to nonsubstituted ones, k is the rate constant, K_i is the adsorption constant of component I, K is the equilibrium constant, and n is the exponent of inhibition term. The value $n = 3$ for HDS indicates that the underlying assumption of the rate-determining step for both types of reaction is the surface reaction between adsorbed reactants and two competitively adsorbed hydrogen atoms.

As Froment and Bischoff (1990) pointed out, the concept of a rate-determining step is not an essential restriction of LHHW rate equations, but it seems to be necessary to apply the more general approach in very few kinetic studies. However, an essential characteristic of LHHW is to account explicitly for the interaction of the reacting components with the catalytic surface. The assumptions behind Langmuir's equation have often been recalled, and the LHHW approach is thus still considered only a systematic, semiempirical formalism to catalytic kinetics, but its basic restrictions generally lead to deviations that are minor with respect to the inaccuracies associated with kinetic experimentation. Boudart (1986) pointed out the validity of this approach. When the catalytic surface is almost completely covered by species, the nonuniformities are no longer felt. In such a case, the LHHW rate equations, based on the Langmuir isotherm, are not only useful but also correct. In all cases, the LHHW formalism provides physical intuition, improvable rate equations, and mechanistic insight unattainable through empirical rate laws.

Examples of applying the LHHW formalism to the kinetic rate laws will be further exploited with the pathway-level detailed kinetic modeling of naphtha reforming (Chapter 7), heavy paraffin hydrocracking (Chapter 8), naphtha hydrotreating (Chapter 9), and gas oil hydroprocessing (Chapter 10).

4.2.2 KINETIC RATE LAWS AT THE MECHANISTIC LEVEL

At the mechanistic level, the kinetic rate laws are straightforward. In principle, each reaction is an elementary step of the fundamental reaction mechanism. The rate law is just a simple mass action rate. For example, for the elementary reaction $A + B \rightarrow R + S$, the rate law is simply

$$r = k\,(C_A\,C_B - C_R\,C_s/K) \qquad (4.8)$$

where k is the rate constant, K is the reaction equilibrium constant, and C_i is the concentration of species i.

For homogeneous systems, the rate law at the mechanistic level is almost always first or second order. Since each reaction pathway is a simple combination of several mechanistic reaction steps, the rate law at the pathways level is a simplification of the rate law at the mechanistic level. For example, consider at the simple overall reaction $A \rightarrow B + C$ in free radical chemistry. The corresponding Rice–Herzfeld mechanism is

$$A \xrightarrow{\;\alpha\;} \beta$$

$$\beta + A \xrightarrow{\;\text{II}\;} \mu + B$$

$$\mu \xrightarrow{\;\text{I}\;} \beta + C \qquad (4.9)$$

$$2\beta, 2\mu, \beta + \mu \xrightarrow{\;\omega\;} T.P.$$

The overall rate derived from the mechanism in Equation 4.9 is

$$\mathbf{r}_A = \frac{\left(\alpha k_{II}^2 A / \omega\right)^{1/2} A}{\sqrt{1 + \gamma\left(\frac{k_{II}A}{k_I}\right) + \gamma'\left(\frac{k_{II}A}{k_I}\right)^2}} \tag{4.10}$$

With the *statistical termination* approximation, Equation 4.10 is reduced to

$$\mathbf{r}_A = \frac{\left(\alpha k_{II}^2 A / \omega\right)^{1/2} A}{1 + k_{II}A/k_I} \tag{4.11}$$

Obviously, the rate law derived from the mechanism is much more complex and fundamental than just the simple empirical mass action rate law at the pathways level. Most kinetic modeling work of homogeneous chemistry like pyrolysis is already in the mechanistic level and is thus the most popular choice.

For heterogeneous catalyzed systems, in principle, the rate law at the mechanistic level is equivalent to the LHHW formalism at the pathways level to describe the kinetics. For example, for the simple reaction $A \rightarrow B$ at the pathways level, the corresponding reaction mechanism includes the chemisorption, surface reaction, and desorption:

$$A + l \Leftrightarrow Al$$

$$Al \Leftrightarrow Bl \tag{4.12}$$

$$Bl \Leftrightarrow B + l$$

The overall rate derived from the above mechanism (Aris, 1965) is

$$\mathbf{r}_A = \frac{l_0(A - B/K)}{\left(\frac{1}{K_A k_{sr}} + \frac{1}{k_A} + \frac{1}{Kk_B}\right) + \left(\frac{1}{K_A k_{sr}} + \frac{1+K_{sr}}{Kk_B}\right)K_A A + \left(\frac{1}{K_A k_{sr}} + \frac{1+K_{sr}}{Kk_A}\right)K_B B} \tag{4.13}$$

When assuming the adsorption control, Equation 4.13 can be reduced to

$$\mathbf{r}_A = \frac{l_0 k_A(A - B/K)}{1 + \frac{K_A}{K} B + K_B B} \tag{4.14}$$

When assuming the surface reaction control, Equation 4.13 can be reduced to

$$\mathbf{r}_A = \frac{l_0 k_{sr} K_A (A - B/K)}{1 + K_A A + K_B B} \tag{4.15}$$

When assuming the desorption control, Equation 4.13 can be reduced to

$$\mathbf{r}_A = \frac{l_0 k_B K (A - B/K)}{1 + K_A A + K K_B A} \tag{4.16}$$

Equation 4.14 to Equation 4.16 can be derived directly from Table 4.1. This simple example shows that a simple rate law at the mechanistic level is equivalent to the complex LHHW formalism without the rate-determining step assumption. However, if we construct the rate law from Table 4.1, the underlying rate-determining step has to be assumed and justified.

For a heterogeneously catalyzed process, both the mechanistic modeling and pathways modeling have advantages and disadvantages. The mechanistic model describes the process at a more fundamental level than the pathways model, so the rate parameters are more fundamental and applicable. In terms of rate law, there are fewer *a priori* assumptions, such as the rate-determining step assumption, in mechanistic level models than in the LHHW formalism for pathways level modeling. The trade-off is that the mechanistic models are much larger than the corresponding pathways level models because of the explicit accounting of all the intermediates. Therefore, both the formulation and solution of mechanistic models takes more CPU time and memory than those of pathways level models. To judiciously choose the right level of the kinetic modeling for the complex process, the right balance of the fundamental applicability and real needs has to be considered. There is no doubt that, with the advancement of computer and analytical technologies, kinetic modeling will push more toward the mechanistic level.

In this section, we have briefly reviewed the rate laws for both homogenous and heterogeneous processes at both pathways and mechanistic level modeling. After the rate law has been constructed and implemented, the next step is how to organize and evaluate the associated rate parameters including rate constants, k, equilibrium constants, K, and adsorption constants, K_i, for various reactions and components. The LFER concept that will be discussed in the following section sets a path forward.

4.3 OVERVIEW OF LINEAR FREE ENERGY RELATIONSHIPS

Korre's analysis (1995) of LFERs is summarized here. Methods for the prediction of reactivity from structure have long been sought in physical organic chemistry. In principle, such attempts stem directly from the thermodynamic formulation of the equilibrium and rate constants:

$$K_{eq} = e^{-\frac{\Delta G}{RT}} = e^{\frac{\Delta S}{R} - \frac{\Delta H}{RT}} \tag{4.17}$$

$$k = \frac{k_B T}{h} e^{-\frac{\Delta G^{\ddagger}}{RT}} = \frac{k_B T}{h} e^{\frac{\Delta S^{\ddagger}}{R} - \frac{\Delta H^{\ddagger}}{RT}} \tag{4.18}$$

where K_{eq} is the equilibrium constant; ΔG, ΔS, and ΔH are the free energy, entropy, and enthalpy differences between products and reactants; R is the ideal gas constant; T is the absolute temperature; k is the rate constant; k_B and h are the Boltzmann and Planck constants, respectively; and $\Delta G^{\ddagger}$, $\Delta S^{\ddagger}$, and $\Delta H^{\ddagger}$ are the free energy, entropy, and enthalpy differences between transition state complexes and reactants.

It follows from Equation 4.17 and Equation 4.18 that estimation of free energies of reaction and activation (ΔG and $\Delta G^{\ddagger}$) would allow estimation of equilibrium and rate constants. This is a considerable task. Even if the enthalpies of reaction and activation were readily available, the corresponding entropies are difficult to measure or calculate. The traditional way around that impasse has been to focus not on absolute values of free energies of reaction and activation, but on establishing the rules along which these quantities vary within a series of analogous reactions. This led to the concept of a reaction family.

The LFER concept exploits the systematic differences in Equation 4.17 and Equation 4.18 between members of a reaction family. In a reaction family, a homologous series of reactants is subject to the same reaction. A classic example is the hydrolysis of substituted benzoic acids (Hammett, 1937). Within the Hammett paradigm, equilibrium and rate constants can be defined in a relative manner, from one member of the family to another:

$$\frac{K_{eq}^i}{K_{eq}^0} = e^{-\frac{\Delta G_i - \Delta G_0}{RT}} = e^{\frac{\Delta S_i - \Delta S_0}{R} - \frac{\Delta H_i - \Delta H_0}{RT}} \Rightarrow$$

$$\ln K_{eq}^i = \ln K_{eq}^0 + \frac{\Delta(\Delta S_{i-0})}{R} - \frac{\Delta(\Delta H_{i-0})}{RT} \tag{4.19}$$

$$\frac{k_i}{k_0} = e^{-\frac{\Delta(\Delta G_{i-0}^{\ddagger})}{RT}} = e^{\frac{\Delta(\Delta S_{i-0}^{\ddagger})}{R} - \frac{\Delta(\Delta H_{i-0}^{\ddagger})}{RT}} \Rightarrow$$

$$\ln k_i = \ln k_0 + \frac{\Delta(\Delta S_{i-0}^{\ddagger})}{R} - \frac{\Delta(\Delta H_{i-0}^{\ddagger})}{RT} \tag{4.20}$$

In Equation 4.19 and Equation 4.20, subscript ($_0$) refers to an arbitrary reference reaction, while subscript ($_i$) refers to any other reaction in the family. The quantities $\Delta(\Delta G_{i-0})$, $\Delta(\Delta S_{i-0})$, and $\Delta(\Delta H_{i-0})$ refer to the free energy, entropy, and enthalpy of reaction differences between reaction ($_i$) and reaction ($_0$). The quantities $\Delta(\Delta G^{\ddagger}_{i-0})$, $\Delta(\Delta S^{\ddagger}_{i-0})$, and $\Delta(\Delta H^{\ddagger}_{i-0})$ refer to the free energy, entropy, and enthalpy of activation differences between reaction ($_i$) and reaction ($_0$).

Equation 4.19 and Equation 4.20 imply that the equilibrium and rate constants can be correlated in terms of entropy and enthalpy differences. This invites further simplification for the special case of the reaction family. Implicit in the reaction family concept is the idea that as long as the reaction center is spatially separated from the substituents (other groups, heteroatoms, alkyl chains), the sterics of the transition state will not change significantly. Rather, the substituents affect the energetics of activation (for example, via charge stabilization, in the case of hydrolysis of substituted benzoic acids). If this is the case, the activation entropy difference $\Delta(\Delta S^{\ddagger}_{i-0})$ in Equation 4.20 can be considered negligible, so that the rate constant k_i is a linear function of rate constant $\ln(k_i)$ is a linear function of only the activation enthalpy difference $\Delta(\Delta H^{\ddagger}_{i-0})$ (Equation 4.21) (Hammett, 1937; Dewar, 1969). This is a very important conclusion, since reaction and activation enthalpies are more accessible through experimental measurements as well as calculations than the corresponding entropies, and it forms the LFER premise.

To summarize, if a reaction family exists, and reactions ($_i$) and ($_0$) conform to it, the following statements are equivalent:

$$\Delta(\Delta S^{\ne}_{i-0}) = 0,$$

$$\frac{A_i}{A_0} = 1,$$

$$\ln\frac{k_i}{k_0} = -\frac{\Delta\left(\Delta H^{\ne}_{i-0}\right)}{RT} = \frac{E^*_0 + RT - E^*_i - RT}{RT} = \frac{\Delta\left(E^*_{0-i}\right)}{RT} \qquad (4.21)$$

where A_i, and E^*_i refer to Arrhenius factors and activation energies, respectively.

The approach is more forgiving than the equalities of Equation 4.21 suggest. An analogous result may be reached if similar factors affect both activation quantities, so that the activation entropy is a linear function of the activation enthalpy (compensation effect, Equation 4.22):

$$\Delta(\Delta S^{\ne}_{i-0}) = A + B \cdot \Delta(\Delta H^{\ne}_{i-0}),$$

$$\ln\frac{k_i}{k_0} = \frac{AT + (BT - 1) \cdot \Delta(\Delta H^{\ne}_{i-0})}{RT} = \frac{A}{R} + \left(\frac{B}{R} - \frac{1}{RT}\right) \cdot \Delta(E^*_{i-0}) \qquad (4.22)$$

In this case, a linear correlation can also be retrieved, but interpretation of the slopes and intercepts is less straightforward than in Equation 4.21.

In either case, direct or indirect estimation of differences in activation enthalpies from one member of the reaction family to another is necessary. Direct estimation can be accomplished either experimentally or computationally. Full calculation of the complete energy surface for reacting systems is possible with reasonable computational demands for small systems; if the current trend in increasing availability of computational power continues, larger systems will be easily treated as well. In the meanwhile, indirect estimations from structural properties of the reactants, intermediates, or products will have to suffice.

A classic methodology for activation enthalpy estimation involves the Evans–Polanyi principle. Following the experimental observation that the activation energies of many reactions in the gas phase linearly correlate with the enthalpy change during the reaction (Equation 4.23), geometrical arguments were enlisted to justify this observation theoretically. As a result, the rate constants in a reaction family can be expressed as a function of the enthalpies of reaction, which are more easily accessible since they are differences between heats of formation of stable compounds.

$$E_i^* = C + D \cdot \Delta H_{rxn} \tag{4.23}$$

$$\ln \frac{k_i}{k_0} = \frac{\Delta\left(E_{0-i}^*\right)}{RT} = D \cdot \frac{\Delta\left(\Delta H_{rxn}^{0-i}\right)}{RT}, \quad \text{or} \tag{4.24}$$

$$\ln \frac{k_i}{k_0} = \frac{A}{R} + \left(\frac{B}{R} - \frac{1}{RT}\right) \cdot D \cdot \frac{\Delta\left(\Delta H_{rxn}^{i-0}\right)}{RT} \tag{4.25}$$

An elegant overview of this classical treatment can be found elsewhere (Dewar, 1969).

Although a large number of correlations have been formulated over the years for a very diverse set of reaction families, LFERs in the strict Hammett sense cannot always be justified. More generally, reactivity has often been semiempirically described in terms of reactivity indices (RIs), which are a measure of the activation enthalpies and are pertinent to each reaction:

$$\ln k_i = a + b * RI_i \tag{4.1}$$

A RI will often correlate linearly with ΔH_{rxn} and $\Delta H^{\ddagger}$ over a useful range of reaction family members. The value to reaction engineering modeling is that the constants a and b can be determined from a minimum basis set of model compound information via experiments specifically designed for that purpose. The semiempirical QSRCs, such as Equation 4.1, can then be used to predict rate constants for reactions in complex mixtures, given the reactivity index, RI_i, derived from the molecular structure.

RIs are generally electronic or energetic properties associated with the molecular structure of the reactants, products, or intermediates that indicate potential for reactivity. These indicators are usually directly tied to the overall molecular structure (such as heat of formation, ionization potential, or electron potential) or may be tied to reaction sites within a molecule (such as atomic charge or bond order). In either case, these parameters combine important electronic or energetic features of the complex reaction coordinate into a simple single value for molecular indices or a series of values for site-based indices. The rate at which different molecules react (kinetics) or the potential reactivity of a particular site (selectivity) can be estimated by computing the molecular and site-based RIs. The challenge lies in determining an appropriate index for each reaction family.

In principle, the RI is tied to the governing reaction mechanism. Early transition states may be viewed as small perturbations of the reactant structure, and therefore, electronic properties that characterize the electronic structure of the reactant may be suitable choices for RIs, such as the electron density (total, π, and spin), atomic charge, free valency, atom polarizability, bond order, dipole moment, and electrostatic potential. For the late transition-state complexes, a significant change in the electronic structure occurs before crossing the reaction barrier, and the complex assumes the form of a perturbation on the products or stable intermediates. The kinetics for these systems are often described by RIs that reflect the changes attributable to the overall energies of the reaction or formation of intermediates, such as the heat of reaction, proton affinity, localization, and Dewar number. Intermediate transition states contain elements of the electronic structure of both the reactants and products and may be described by a charge-transfer type complex. The possible RIs can be Brown's localization energy and the Fukui superlocalizability.

Although correlation analysis is a largely successful field in homogeneous chemistry, QSRCs have not yet been used as extensively in heterogeneous catalysis because of uncertainty about the elementary steps (Dunn, 1968). Nevertheless, their applicability as a modeling tool is well established. Mochida and Yoneda (1967a,b,c) demonstrated the relevance of QSRCs to heterogeneous reactions. Their work focused on carbenium ion chemistry, particularly dealkylation of alkylbenzenes. Landau and coworkers explored this subject more recently (Landau, 1991; Landau et al., 1992). Other researchers have pondered the subject of QSRC development on catalytic hydroprocessing of substituted benzenes, phenols, and heteroaromatic compounds (Kochloefl and Bazant, 1967; Aubert et al., 1988; Moreau et al., 1988, 1990) and on zeolite-catalyzed acylation of aromatics (Chiche et al., 1987) and have formulated several Hammett-type correlations. More recently, Neurock and coworkers (Neurock, 1992; Neurock and Klein, 1993a) revisited both acid-catalyzed and metal-catalyzed reactions with new input from computational quantum chemistry. Korre (1995) and Russell (1992) systematically studied the LFER correlations in the catalytic hydrocracking of unsubstituted PNAs and alkyl-PNAs including alkylbenzenes.

4.4   REPRESENTATIVE RESULTS AND SUMMARY OF LFERS FOR CATALYTIC HYDROCRACKING

To demonstrate the wide range of applicability of the LFER concept, a series of representative results developed for the study of catalytic hydrocracking are presented here that will be used for the industrial applications development in Part II of this book. The LFER formats should be applicable to a wide range of metal-catalyzed and acid-catalyzed chemistries because the fundamental reaction families including adsorption, hydrogenation, isomerization, ring opening, and dealkylation are common to any hydrocarbon conversion chemistry.

Figure 4.1 (Neurock, 1992) shows the LFER correlation used to estimate adsorption constants. The experimental adsorption coefficients were determined through experiments where competitive inhibition data were represented by the LHHW adsorption constant K_i (La Vopa and Satterfield, 1988; Neurock and Klein, 1993a). The proton affinity (PA) was taken as a measure of gas-phase basicity and computed according to the thermochemical cycle: $M + H^+ \rightarrow MH^+$. Thus, the PA is given by

$$PA = -\Delta H_{rxn} = \Delta H_f^0(M) + \Delta H_f^0(H^+) - \Delta H_f^0(MH^+)$$

and can be computed using MNDO (modified neglect of diatomic overlap) calculations. As we can see from Figure 4.1, the MNDO calculated PAs correlate linearly with the experimental adsorption parameters.

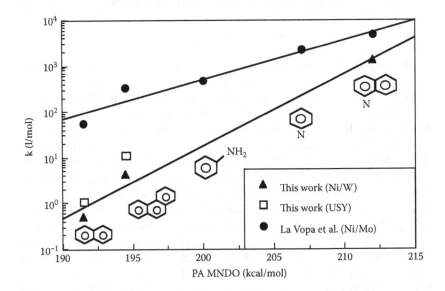

FIGURE 4.1 LFER correlation of proton affinity (PA) used to estimate the adsorption constants K_i.

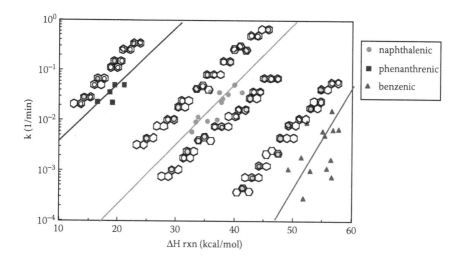

FIGURE 4.2 LFER correlations of experimental hydrogenation rate constants and heat of reaction for aromatic compounds.

This good correlation shows the organizational capability of LFERs, which suggests an efficient methodology for the estimation of kinetic parameters for many similar compounds that have not been studied experimentally. For a similar molecule, this would simply require a PA calculation. From the chemical significance perspective, the correlation between K_i and PA points to an acid–base interaction between the catalyst and reactant (La Vopa and Satterfield, 1988).

Figure 4.2 (Korre, 1995) shows very good correlations of the experimental hydrogenation rate constants with the heat of reaction as the RI for various aromatics, including benzenic, naphthalenic, and phenanthrenic compounds. The highlighted rings denote the site of hydrogenation. Although the hydrogenation of aromatic compounds seems to provide special challenges of accounting for the multiplicity of the reaction sites per molecule, the overall heat of reaction has served as a very good RI to characterize these reactions.

Figure 4.3 shows the LFER correlation of the kinetics of acid-center transformation reaction families (both isomerization and ring opening), with the heat of formation of the carbenium ion intermediate as the RI. The correlation was successful. The isomerization and ring opening share the similar underlying reaction mechanism and have the same slope in Figure 4.3. However, it is clear that the isomerization reaction is much slower than the ring opening reaction.

Figure 4.4 shows Mochida and Yoneda's (1967) classical example of LFER correlations on the dealkylation rate parameters, with the stability of the alkyl ion based on the carbenium ion chemistry reaction mechanism. Neurock and Klein (1993) also used the heat of reaction of the rate-determining step based on the carbenium ion chemistry reaction mechanism as a RI and successfully correlated it with the dealkylation kinetics.

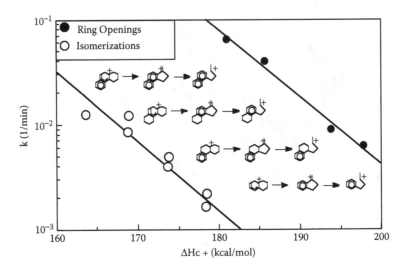

FIGURE 4.3 LFER correlation of the experimental isomerization and ring opening rate constants with the stability of the carbenium ion intermediates.

The above examples qualitatively demonstrate the feasibility of using the LFER concept as a reaction engineering tool to organize rate parameters. Table 4.2 summarizes a comprehensive set of quantitative correlations and LFER parameters developed in the past few years by Klein and coworkers on the catalytic hydrocracking of various aromatics, PNAs (Korre, 1995), alkylbenzenes, and

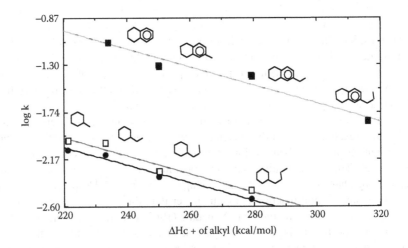

FIGURE 4.4 LFER correlation of experimental dealkylation rate constant with the heat of formation of the intermediate carbenium alkyl ion (Data from Mochida and Yoneda, 1967a,b, and c).

TABLE 4.2
Linear Free Energy Relationships in Catalytic Hydrocracking

Compound Class	Reaction Families	$\ln(K) = a + b*RI_1 + c*RI_2$			
		a	b	c	RI
PNAs (1–4 aromatic rings)	Hydrogenation				
	Adsorption $\ln(K_1)$	1.324	0.887/RT	0.123/RT	RI_1 : N_{AR} (no. of aromatic rings) RI_2 : N_{SC} (no. of saturated C atoms)
	Equilibrium $\ln(K_{eq})$	—	−7.017	1/RT−0.54707	RI_1 : n (hydrogen stoichiometry) RI_2 : $|\Delta H_r^0|$ (heat of reaction at 25°C)
	Reaction $\ln(k_{sr})^*$	−6.613	−5.767	−0.228/RT	RI_1 : n (hydrogen stoichiometry) RI_2 : $|\Delta H_r^0|$ (heat of reaction at 25°C)
	Acid center transformation				
	Adsorption $\ln(K_1)$	0.182	1.934/RT	0.187/RT	RI_1 : N_{AR} (no. of aromatic rings) RI_2 : N_{SC} (no. of saturated C atoms)
	Equilibrium $\ln(K_{eq})$	−1.775	1.469−1/RT	—	RI_1 : ΔH_r° (heat of reaction at 25°C)
	Isomerization $\ln(k_i)$	−181.32 + 305/RT	1−1.709/RT	—	RI_1 : $\Delta H_{f,C}^+$ (heat of formation of carbenium ion)
	Ring opening $\ln(k_{ro})^{**}$	−163.18 + 271/RT	1−1.709/RT	—	RI_1 : $\Delta H_{f,C}^+$ (heat of formation of carbenium ion)

(Continued)

TABLE 4.2(Continued)
Linear Free Energy Relationships in Catalytic Hydrocracking

Compund Class	Reaction Families	$\ln(K) = a + b * RI_1 + c*RI_2$			RI
		a	b	c	
Alkyl PNAs (1–4 aromatic rings)	Hydrogenation				
	ln(k) phenanthrenic	−14.017	0.47072	—	RI_1 : ΔH_r (heat of reaction)
	ln(k) naphthalenic	−19.608	0.40913	—	RI_1 : ΔH_r (heat of reaction)
	Ring dealkylation	1.6274	−0.17044	—	RI_1 : ΔH_r (heat of reaction)
	ln(k) (side chain length ~C_{8-12})				
	Ring closure	−0.48669	−0.04103	—	RI_1 : $\Delta H_{f,C}{}^+$ (heat of formation of carbenium ion)
	ln(k)				
Alkylbenzenes	Ring dealkylation	1.5268	−0.03333	—	RI_1 : $\Delta H_{f,C}{}^+$ (heat of formation of carbenium ion)
	ln(k)	15.291	−0.49444	—	RI_1 : ΔH_r (heat of reaction)
	(side chain length ~C_2-C_7)				
	Ring closure	7.6899	−0.08364	—	RI_1 : $\Delta H_{f,C}{}^+$ (heat of formation of carbenium ion)
	ln(k) tetralin deriv.	−1.9401	−0.05042	—	RI_1 : ΔH_r (heat of reaction)
	ln(k) indan deriv.	4.7293	−0.06575	—	RI_1 : $\Delta H_{f,C}{}^+$ (heat of formation of carbenium ion)
		21.625	−0.4128	—	RI_1 : ΔH_r (heat of reaction)

Note: * : $k = k_{st} / (K_i P_{H2}{}^n)$; ** : $k = k_{ro} / P_{HV}$; ammonia inhibition: $K_H = 407$ l/mol, $K_A = 2627$ l/mol at 350°C alkyl-PNAs (Russell, 1992). These correlations will be used in the hydrocracking and hydroprocessing kinetic models discussed in Part II.

The qualitative and quantitative results shown here clearly demonstrate that the LFER is a very useful organization and estimation tool for the summary of experimental data in complex reaction systems. These relationships provide direct links between structure-engendered reactivity properties and experimental kinetic parameters, allowing for the reduction of thousands of kinetic parameters into a manageable $O(10)$ series of slopes and intercepts. LFERs provide the means for organizing existing experimental data as well as presenting a basis for predicting the kinetics of many experimentally unstudied systems.

4.5 SUMMARY AND CONCLUSIONS

In this chapter, we have focused on the methodologies of organizing and evaluating rate parameters and rate laws in the development of detailed kinetic models.

The rate laws of the complex reaction network are reviewed at both the pathways level and mechanistic level. At the pathways level, for homogeneous systems, the rate law formula is a simple mass action rate as long as the reaction order is determined via experiments; for heterogeneous systems, for example, all the catalyzed processes, the LHHW formalism is still frequently used. At the mechanistic level, the rate law is just as simple as the mass action rate for each elementary step.

The LFER concept as a means of organizing kinetic data is reviewed from the classical transition state theory. The reaction family concept is introduced along the way. Some representative results developed for catalytic hydrocracking, including adsorption, hydrogenation, isomerization, ring opening, and dealkylation, are presented to demonstrate the wide range of applicability of the LFER concept. A comprehensive set of LFERs developed for catalytic hydrocracking was organized and tabulated for easy reference.

REFERENCES

Andreu, P. and Ramirez, M.M., The influence of reactant structure on reactivity and isotope effects in the cracking of alkylbenzenes over an acid catalyst, *Proc. 6th Int. Congr. Catal.*, 1, 593–602, 1977.

Aris, R., *Introduction to the analysis of Chemical Reactors,* Prentice-Hall, Englewood Cliffs, NJ, 1965.

Aubert, C., Durand, R., Geneste, P., and Moreau, C., Factors affecting the hydrogenation of substituted benzenes and phenols over a sulfided $NiO\text{-}MoO_3/\gamma\text{-}Al_2O_3$ catalyst, *J. Catal.*, 112, 12–20, 1988.

Baltanas, M.A., Van Raemdonck, K.K., Froment, G.F., and Mohedas, S.R., Fundamental kinetic modeling of hydroisomerization and hydrocracking on noble-metal-loaded faujacites. 1. Rate parameters for hydroisomerization, *Ind. Eng. Chem. Res.,* 28, 899–910, 1989.

Benson, S.W., *Thermochemical Kinetics,* 2nd ed., John Wiley & Sons, New York, 1976.

Boudart, M., Classical catalytic kinetics: a placebo or the real thing? *Ind. Eng. Chem. Fundam.,* 25, 656, 1986.

Chiche, B., Finiels, A., Gauthier, C., and Geneste, P., The effect of structure on reactivity in zeolite catalyzed acylation of aromatic compounds: a ρ-σ+ relationship, *Appl. Catal.,* 30, 365–369, 1987.

Corma, A., Miguel, P.J., Orchilles, A.V., and Koermer, G.S., Cracking of long-chain alkyl aromatics on USY zeolite catalysts, *J. Catal.,* 135, 45–59 1992.

Dewar, M.J.S., *The Molecular Orbital Theory of Organic Chemistry,* McGraw-Hill, New York, 1969.

Dunn, I.J., Linear free energy relationships in modeling heterogeneous catalytic reactions, *J. Catal.,* 12, 335–340, 1968.

Evans, M.G. and Polanyi, M., Inertia and driving force of chemical reactions, *Trans. Faraday Soc.,* 34, 1138, 1938.

Froment, G.F., Kinetics of the hydroisomerization and hydrocracking of paraffins on a platinum containing bifunctional Y-zeolite, *Catal. Today,* 1, 455–473, 1987.

Froment, G.F. and Bischoff, K.B., *Chemical Reactor Analysis and Design,* 2nd ed., John Wiley & Sons, New York, 1990.

Gray, M.R., Lumped kinetics of structural groups: hydrotreating of heavy distillate, *Ind. Eng. Chem. Res.,* 29, 505–512, 1990.

Hammett, L.P., The effect of structure upon the reactions of organic compounds. Benezene., *J. Am. Chem. Soc.,* 59, 96, 1937.

Haynes, H.W.J., Parcher, J.F., and Helmer, N.E., Hydrocracking polycyclic hydrocarbons over a dual-functional zeolite (faujacite)-based catalyst, *Ind. Eng. Chem. Process Des. Dev.,* 22, 401–409, 1983.

Hoek, A., Huizinga, T., Esener, A.A., et al., New catalyst improves heavy feedstock hydrocracking, *Oil Gas J.,* 77–82, 89(16), 1991.

Jik, K.J., Uchytil, J., and Kraus, M., Hydrodealkylation of alkylbenzenes on a nickel-molybdenum/alumina catalyst, *Appl. Catal.,* 35, 289–298, 1987.

Kochloefl, K. and Bazant, V., Hydrogenolysis of saturated hydrocarbons on a nickel catalyst I. Kinetics of hydrogenolysis of ethylcyclohexane and reactivity of alkylcyclohexanes, *J. Catal.,* 8, 250–260, 1967.

Korre, S., Quantitative Structure/Reactivity Correlations as a Reaction Engineering Tool: Applications to Hydrocracking of Polynuclear Aromatics, Ph.D. thesis, University of Delaware, Newark, 1995.

Landau, R.N., Chemical Modeling of the Hydroprocessing of Heavy Oil Feedstocks, Ph.D. thesis, University of Delaware, 1991.

Landau, R.N., Korre, S.C., Neurock, M.N., Klein, M.T., and Quann, R.J., Hydrocracking of heavy oils: development of structure/reactivity correlations for kinetics, *Am. Chem. Soc. Div. Fuel Chem. Prepr.,* 37, 1871, 1992.

Lapinas, A.T., Klein, M.T., Gates, B.C., Macris, A., and Lyons, J.E., Catalytic hydrogenation and hydrocracking of fluoranthene: reaction pathways and kinetics, *Ind. Eng. Chem. Res.,* 26, 1026–1033, 1987.

Lapinas, A.T., Klein, M.T., Gates, B.C., Macris, A., and Lyons, J.E., Catalytic hydrogenation and hydrocracking of fluorene: reaction pathways, kinetics and mechanisms, *Ind. Eng. Chem. Res.,* 30, 42–50, 1991.

La Vopa, V. and Satterfield, C.N., Poisoning of thiophene hydrodesulfurization by nitrogen compounds, *J. Catal.,* 110, 375, 1988.

Lemberton, J.-L. and Guisnet, M., Phenanthrene hydroconversion as a potential test reaction for the hydrogenating and cracking properties of coal hydroliquefaction catalysts, *Appl. Catal.,* 13, 181–192, 1984.

Liguras, D.K. and Allen, D.T., Structural models for catalytic cracking. 1. Model compound reactions, *Ind. Eng. Chem. Res.,* 28, 665–673, 1989.

Machácek, H., Kochloefl, K., and Kraus, M., Catalytic dealkylation of alkylaromatic compounds. XV. The effect of structure of alkylnaphthalenes on the rate of their hydrodealkylation on a nickel catalyst, *Collection Czechoslov. Chem. Commun.,* 31, 576–584, 1966.

McLafferty, F.W., *Interpretation of Mass Spectra,* University Science Books, Mill Valley, CA, 1980.

Mochida, I. and Yoneda, Y., Linear free energy relationships in heterogeneous catalysis. I. Dealkylation of alkylbenzenes on cracking catalysts., *J. Catal.,* 7, 386–392, 1967a.

Mochida, I. and Yoneda, Y., Linear free energy relationships in heterogeneous catalysis. II. Dealkylation and isomerization reactions on various solid acid catalysts, *J. Catal.,* 7, 393–396, 1967b.

Mochida, I. and Yoneda, Y., Linear free energy relationships in heterogeneous catalysis. III. Temperature effects in dealkylation of alkylbenzenes on the cracking catalysts, *J. Catal.,* 8, 223–230, 1967c.

Moreau, C., Aubert, C., Durand, R., Zmimita, N., and Geneste, P., Structure-activity relationships in hydroprocessing of aromatic and heteroaromatic model compounds over sulphided NiO-MoO$_3$/γ-Al$_2$O$_3$ and NiO-WO$_3$/γ-Al$_2$O$_3$ catalysts: chemical evidence for the existence of two types of catalytic sites, *Catal. Today,* 4, 117–132, 1988.

Moreau, C., Joffre, J., Saenz, C., and Geneste, P., Hydroprocessing of substituted benzenes over a sulfided CoO-MoO$_3$/γ-Al$_2$O$_3$ catalyst, *J. Catal.,* 122, 448–451, 1990.

Neurock, M.T., A Computational Chemical Reaction Engineering Analysis of Complex Heavy Hydrocarbon Reaction Systems, Ph.D. thesis, University of Delaware, Newark, 1992.

Neurock, M. and Klein, M.T., Linear free energy relationships in kinetic analyses: applications of quantum chemistry, *Polyc. Arom. Compounds,* 3, 231–246, 1993a.

Neurock, M. and Klein, M.T., When you can't measure—model, *Chem. Tech,* 23, 26–32, 1993b.

Quann, R.J. and Jaffe, S.B., Structure-oriented lumping: describing the chemistry of complex hydrocarbon mixtures, *Ind. Eng. Chem. Res.,* 31, 2483–2497, 1992.

Russell, C.L., Hydrocracking Reaction Pathways, Kinetics and Mechanisms of n-Alkylbenzenes, M.S. thesis, University of Delaware, 1992.

Sapre, A.V., Reaction Networks and Kinetics in High Pressure Hydrodesulfurization and Hydrogenation Catalyzed by Sulfided CoO-MoO$_3$/γ-Al$_2$O$_3$, Ph.D. thesis, University of Delaware, 1980.

Steijns, M. and Froment, G.F., Hydroisomerization and hydrocracking. 3. Kinetic analysis of rate data for n-decane and n-dodecane, *Ind. Eng. Chem. Prod. Res. Dev.,* 20, 660–668, 1981.

Sullivan, R.F., Boduszynski, M.M., and Fetzer, J.F., Molecular transformations in hydrotreating and hydrocracking, *Energy & Fuels*, 3, 603–612, 1989.

Sullivan, R.F., Egan, C.J., and Langlois, G.E., Hydrocracking of alkylbenzenes and polycyclic aromatic hydrocarbons on acidic catalysts: evidence for cyclization of the side chains, *J. Catal.*, 3, 183–195, 1964.

Trauth, D.M., Structure of Complex Mixtures through Characterization, Reaction and Modeling, Ph.D. thesis, University of Delaware, Newark, 1993.

Yang, K.H. and Hougen, O.A., Determination of Mechanism of catalyzed gaseous reactions, *Chem. Eng. Progress*, 46/37, 146-147, 1950.

5 Matching the Equation Solver to the Kinetic Model Type

5.1 INTRODUCTION

Detailed kinetic models (DKMs) have evolved considerably over the past two decades, in several cases becoming the kinetics kernels of complex process models for industrial applications. In the previous chapters, we discussed in detail how to construct automatically a rigorous and useful reaction network for complex process chemistries (Chapter 3) and how to organize and evaluate rate constants for each reaction in the reaction network (Chapter 4). In the context of the complete model building work process as discussed in Chapter 1, the model building stage is now fully automated and thus is no longer the rate-determining step; often, the model tuning process, which involves repetitive solution and optimization stages, becomes more time consuming. This is especially so if an off-the-shelf numerical package is used without consideration of the type of kinetic model. In this chapter, we will focus on the fast and efficient solution of detailed kinetic models from both the mathematical and the chemical and physical points of view. We will illustrate how the nature of the chemical problems and reaction kinetics provide insights to improve the model solving techniques and thus speed up the model development process as a whole.

Mathematically, kinetic models can in general be represented as either ordinary differential equation (ODE) systems or differential algebraic equation (DAE) systems, depending on the reactor type. DKMs, due to their detailed nature, have a very large number of species, reactions, and rate parameters associated with the reaction network. Thus, the numerical solutions of DKMs often involve solving large system of DAEs or ODEs, which are often stiff; the associated Jacobian matrices are often very sparse for real problems and hence difficult to solve. All these numerical issues associated with DKMs are reviewed and discussed in Section 5.2. Then, in Section 5.3, a set of experiments are set up on the computer to illustrate the model solving efficiency of various numerical solvers in two case studies: a gas oil hydrocracking model at the pathways level and a C8 hydrocracking model at the mechanistic level. Various scenarios are analyzed, and the results are presented in Section 5.4. Among other issues, the stiffness of the systems is considered, and the Jacobian matrix sparseness is exploited and handled both implicitly and explicitly for DKMs. Model solving efficiency with and without considering stiffness and sparseness is analyzed.

Finally, based on the findings in the experiments, the guidelines for optimal solution methodologies are proposed for solving large DKM systems.

5.2 MATHEMATICAL BACKGROUND

We will first review the underlying numerical methods (Beris, 1998) for solving ODE and DAE systems and then exploit the stiffness nature and sparseness nature embedded in the DKM systems, respectively.

5.2.1 UNDERLYING NUMERICAL METHODS FOR SOLVING DKM SYSTEMS

The kinetic model generally can be described as the following initial value problem:

$$\dot{y} = f(y), y \in R^N$$
$$y(t_0) = y_0$$

(5.1)

More specifically, the ODE system can be written in the form

$$dy/dt = f(t, \mathbf{y})$$

(5.1a)

and the DAE system can be written in the form

$$dy/dt = f(t, \mathbf{y}, \beta);$$

(5.1b)

$$0 = g(t, \mathbf{y}, \beta),$$

(5.1c)

where $\mathbf{y}$ and β are the vectors of dependent variables, and t is the independent variable. In the context of our molecule-based kinetic models, $\mathbf{y}$ and β can be thought of as observable molecules and unobservable intermediates, respectively.

The numerical solution to Equation 5.1 is generated as discrete values y_n at t_n, which obey a linear multistep formula:

$$y_n = \sum_{i=1}^{K_1} \alpha_{n,i} y_{n-i} + h_n \sum_{i=0}^{K_2} \beta_{n,i} \dot{y}_{n-i}$$

(5.2)

where step size $h_n = t_n - t_{n-1}$. For nonstiff problems, the Adams–Moulton method of order q is characterized by $K_1 = 1$ and $K_2 = q - 1$. For stiff problems, the backward differentiation formula (BDF) method of order q has $K_1 = q$ and $K_2 = 0$.

In either the nonstiff or stiff case, the generally nonlinear system

$$G(y_n) \equiv y_n - h_n\beta_{n,0}f(t_n, y_n) - a_n = 0$$

$$where, a_n = \sum_{i>0} (\alpha_{n,i}y_{n-i} + h_n\beta_{n,i}\dot{y}_{n-i})$$

(5.3)

must be solved at each time step. In the nonstiff case, this is usually done with simple functional (or fixed point) iteration. In the stiff case, this is usually done with some variant of the Newton iteration.

The Newton iteration requires the solution of linear systems of the form

$$G(y_{n(m)}) = -M(y_{n(m+1)} - y_{n(m)})$$

(5.4)

where M is an approximation of the Newton matrix $(\mathbf{I}-h\beta_{n,0}\mathbf{J})$, and $\mathbf{J} = \partial f/\partial y$ is the system *Jacobian*. This linear system can be solved by either a direct (e.g., dense, banded, and diagonal) or an iterative (e.g., generalized minimal residual) method.

5.2.2 STIFFNESS IN DKM SYSTEMS

Stiffness is a characteristic of differential equations that arises from models of real systems where interactions take place on more than one timescale. In other words, the problem is stiff if it contains both very rapidly and very slowly decaying terms. Stiffness is local because a problem may be stiff in some intervals and not in others. In the context of DKMs, especially at the mechanistic level, for example, where some transient reactions occur on a timescale of a few microseconds or less, while a slower steady-state reaction takes place on a timescale of a second or more, stiffness is often encountered.

A quantitative measure of stiffness is usually given by the stiffness ratio

$$\text{Stiffness ratio} = \max[-Re(\lambda i)]/\min[-Re(\lambda i)]$$

(5.5)

The number of steps at the order of the stiffness ratio is required by classical methods, for example, the explicit Runge–Kutta method, to satisfy the stability requirements. Therefore, explicit Runge–Kutta or Adams methods are inappropriate for this class of problems since, in order to assure numerical stability, the step size of these methods must be severely constrained. One should suspect stiffness when a problem is very expensive to integrate with standard methods and the cost of integration (or average step size required) seems to be relatively insensitive to the accuracy requested. Stiffness is inevitable if the Jacobian matrix $\mathbf{J} = \partial f/\partial y$ of the differential equation has eigenvalues $\{\lambda i\}$ with real parts that are predominantly negative and also vary widely in magnitude (Radhakrishnan and

Hindmarsh, 1993; Hindmarsh et al., 1997). In a simple example, reactant R goes through I and converts to product P ($k_2 > k_1$):

$$R \xrightarrow{k_1} I \xrightarrow{k_2} P \qquad (5.6)$$

The eigenvalues of this system are k_1 and k_2, so in this case

$$Stiffness\ ratio = \frac{|\lambda_{max}|}{|\lambda_{min}|} = \frac{k_2}{k_1} \qquad (5.7)$$

The more $k_2 > k_1$ differs, the more I is like an intermediate (e.g., the carbenium ion in the mechanistic level model) rather than an observable product; also, the system is becoming much stiffer. This is why the mechanistic-level model is generally stiffer and thus more difficult to solve than the pathways-level model.

Another characteristic, k(A), the condition number, can help understand the DKM system from the mathematical point of view. The condition number reflects the ratio of the relative error in the solution to the relative error in input parameters.

$$k(A) = \| A \| \bullet \| A^{-1} \| \approx |\lambda max| / |\lambda min| \qquad (5.8)$$

This index normally shows a system's "illness," but the value depends on the specific application. Especially for a large system, some equations weights are much higher than others; some variables have much higher values than others (which is the case for the mechanistic-level DKMs); the system will be more difficult to solve and more prone to numerical instability.

No explicit numerical methods possess the stability required for solving a problem involving a stiff system, unless excessively small step sizes are implemented; only implicit numerical methods should be used to solve a stiff system. Implicit methods designed for stiff systems allow much larger step sizes to be used outside the transient region but require more work per step. The BDFs are often used for stiff systems. The cost of these methods is often dominated by the cost of setting up and solving linear algebraic equations at each step. For this reason, when solving large systems of stiff equations, it is important that the linear algebraic techniques exploit any structure that may exist in the problem.

5.2.3 SPARSENESS IN DKM SYSTEMS

A matrix is sparse if many of its coefficients are zero. Generally, we say that a matrix is sparse if there is an advantage in exploiting its zeros. Effective work with sparse matrices requires special numerical algorithms that can take account of the sparseness, special storage techniques, and special programming techniques. These special techniques can lead to the results being numerically acceptable, the storage demands being minimized, and the computation time and costs being minimized. Generally, sparseness technology can be applied when nonzero elements are at least less than 20% (Beris, 1998).

For an ODE/DAE system, implicit numerical methods lead to systems of linear algebraic equations that in turn lead to Jacobian matrix evaluations. In general, the Jacobian matrices for DKMs are very sparse. In a reaction network with hundreds of species and thousands of reactions, every species is normally only related to a few other species and reactions. This leads the Jacobian to be very sparse. For example, in the mechanistic level C8 hydrocracking model we will test in the next section, which has a total of 81 species and 241 reactions, there are 6002 zeros in the Jacobian. Thus, the sparsity is only 8.5% as calculated below.

$$\text{Sparsity} = 1 - 6002/(81 \times 81) = 8.5\% \qquad (5.9)$$

The sparseness of the Jacobian matrix in solving DKMs is an important characteristic that we should take advantage of to improve the model solving performance. As Hindmarsh (1998) pointed out, it is always better to supply the Jacobian explicitly if you can. The finite difference (FD) estimate is always subject to some error and usually costs more to generate. With all the above understandings, we will exploit the established solvers in the industry that utilize sparseness. In the next section, we will evaluate their benefits.

5.3　EXPERIMENTS

5.3.1　CANDIDATE DKMs

We have chosen two representative DKMs to study model solution techniques for the two main categories of DKMs. Both models are represented as ODE systems: (1) a gas oil hydrocracking pathways model (GO-HDC-Path), which represents gas oil hydrocracking at the pathways level and includes 264 species and 833 reactions, and (2) a C8 hydrocracking mechanistic model (C8-HDC-Mech), which represents C8 hydrocracking at the mechanistic level and includes 81 species and 241 reactions. The details of the gas oil hydrocracking model at the pathways level can be found in Chapter 10, and the paraffin hydrocracking model at the mechanistic level can be found in Chapter 8.

5.3.2　CANDIDATE SOLVERS

Many ODE/DAE solvers have been developed over the decades. From this group, we have chosen four representative and established solvers (DASSL, LSODE, LSODES, and LSODA) to evaluate which one and what configurations we should choose to set up for solving what kinds of DKMs.

DASSL (a *differential/algebraic system solver*), developed by Brenan et al. (1989), solves stiff systems of the form F(t,y,y') = 0, y(t0) = y0, and y'(t0) = y0', where F, y, and y' are vectors. DASSL approximates the derivative using BDF methods. The linear systems that arise are solved by LU (Lower-Upper) direct methods. The Jacobian matrix may be dense or have a banded structure and is either computed by finite differences or supplied directly by the user.

DASSL is so widely used for solving DKMs that it has been used as a default solver whether the DKM is finally modeled as a DAE system or as an ODE system. However, there is generally a big difference in efficiency when applying different solvers for different systems; an ODE solver, not a DAE solver, should be used to solve an ODE system. As we pointed out in Section 5.2.1, it has been observed that large DAE systems can pose convergence problems and are more difficult to solve numerically than the corresponding ODE systems. Hence we have formulated the two candidate DKMs in Section 5.3.1 as ODE systems for faster solutions without loss of accuracy. Also, as the authors of DASSL pointed out (Brenan et al., 1989), the finite difference Jacobian calculation is the weakest part of the DASSL code. For these reasons, we have extended the DKM solving methods in our work to the Livermore Solvers for Ordinary Differential Equations (LSODE) and its variants to complement DASSL.

LSODE is the basic solver of the ODEPACK developed by Hindmarsh and coworkers (Radhakrishnan and Hindmarsh, 1993; Hindmarsh et al., 1997) (ODEPACK software can be downloaded at http://www.netlib.org/.). It solves both stiff and nonstiff systems of the form dy/dt = f. In the stiff case, it treats the Jacobian matrix df/dy as either a full or a banded matrix, and as either user supplied or internally approximated by difference quotients. It uses Adams methods (predictor-corrector) in the nonstiff case and BDF methods in the stiff case. The linear systems that arise are solved by direct methods (LU factor/solve). LSODE supersedes the older GEAR and GEARB packages and reflects a complete redesign of the user interface and internal organization, with some algorithmic improvements.

LSODES, written by Hindmarsh with A.H. Sherman (Hindmarsh et al., 1997), solves systems dy/dt = f, and in the stiff case, treats the Jacobian matrix in general sparse form. It determines the sparsity structure on its own (or optionally accepts this information from the user) and uses parts of the Yale Sparse Matrix Package (YSMP) to solve the linear systems that arise.

LSODA, written by Hindmarsh with L. R. Petzold (Hindmarsh et al., 1997), one of the authors of DASSL, solves systems dy/dt = f with a full or banded Jacobian when the problem is stiff, but it automatically selects between nonstiff (Adams) and stiff (BDF) methods. It uses the nonstiff method initially and dynamically monitors data to decide which method to use.

We will compare the model solving efficiency for all four solvers for the above two DKM models. The Jacobian matrix sparseness will be considered in the experiments and handled both implicitly and explicitly for the DKMs. The model solving efficiency, with or without considering sparseness, will be compared. We will thus propose the new solution methodology for solving large DKM systems based on our experimental results.

TABLE 5.1
Method Flags for Candidate Solvers

MF	Stiffness	Method	Jacobian Source	Jacobian Sparseness
10	Nonstiff	Adams	N.A.	N.A.
22	Stiff	BDF	FD*	Full
21	Stiff	BDF	User-supply	Full
222	Stiff	BDF	FD	Sparse
121	Stiff	BDF	User-supply	Sparse
2	Nonstiff→stiff	Adams→BDF	FD	Full

*FD: Finite Difference Approximation

5.3.3 EXPERIMENT SETUP

The two candidate DKM models (GO-HDC-Path and C8-HDC-Mech) are set up to test four candidate solvers (DASSL, LSODE, LSODES, and LSODA) to compare the model solving performance. The models are solved with various method flags (MFs) as summarized in Table 5.1. A complete set of stiffness (either Adams for nonstiff, BDF for stiff, or dynamic Adams and BDF) and sparseness (either user supplied or FD approximation) combinations are to be tested. For the model C8-HDC-Mech, the explicit Jacobian matrix was generated automatically using the developed algorithm and supplied to explore and utilize its sparseness. All the experiments are run on a Pentium Pro 200 running Linux 2.0.32 Kernel with egcs-1.02 compilers (experimental GNU compliers).

5.4 RESULTS AND DISCUSSION

The experimental results of the setup in Section 5.3.3 are summarized in Table 5.2 and Table 5.3 for model GO-HDC-Path and model C8-HDC-Mech, respectively. We will discuss both cases, analyze the improvable areas in the solvers, and propose the model solving guidelines.

TABLE 5.2
Model GO-HDC-Path Solving Performance for Various Methods

Compile Solver	Debugging								Optimized	
	DASSL	LSODE				LSODES		LSODA	LSODES	
MF	22	10	22	21	10	222	121	2	10	222
CPU(s)[a]	13.15	0.09	4.50	N.A.	0.07	0.36		0.11	0.04	0.06

[a]Pentium Pro 200, Linux 2.0.32, egcs-1.02

TABLE 5.3
Model C8-HDC-Mech Solving Performance for Various Methods

Compile Solver	Debugging								Optimized LSODES	
	DASSL	LSODE			LSODES			LSODA		
MF	22	10	22	21	10	222	121	2	222	121
CPU(s)[a]	1.88	b	2.01	1.46	b	0.44	0.30	2.13	0.11	0.09

[a]Pentium Pro 200, Linux 2.0.32, egcs-1.02
[b]Excessive amount of work done, 500 steps expended

5.4.1 PATHWAYS-LEVEL DKM

Table 5.2 summarizes the model GO-HDC-Path solving performance for various methods. First, we can see DASSL's poor performance on such problems compared with the LSODE family solvers. This confirms our presumption that, in general, an ODE solver should be applied to an ODE system, instead of a DAE solver. The results also show that the Adams–Moulton method performs much better than implicit BDF methods, which indicates that the pathways level models are more nonstiff in nature. This makes sense because in pathways level modeling, only observable species are represented in the reaction network, and the difference between the rate constants for different reaction families is not very large, which leads to a nonstiff problem. This is also the major reason why DASSL performs poorly on such problems since it is designed for solving stiff problems. LSODES, which utilized the sparsity structure of the Jacobian matrix, does a better job than LSODE, in both Adams and BDF methods. LSODA uses nonstiff method initially and dynamically changes to a stiff method when needed (which is not needed for this problem). It does a little bit worse than the direct Adams method throughout but much better than the BDF method throughout, which makes sense.

In summary, for the pathways-level DKM in this experiment, the solver performance ranking is

> LSODES with Adams
> > LSODE with Adams
> > LSODA
> >> LSODES with BDF
> >> LSODE with BDF
> >> DASSL

In addition, the compiler optimization was exploited over the debugging (in general, -O over -g when compiling), which can further improve the model solving performance. Based on this work on pathways level DKMs, the optimized LSODES with the Adams method can get a performance gain over classical

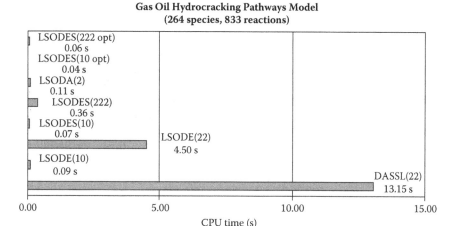

FIGURE 5.1 GO-HDC-Path model solving performance for various solvers.

DASSL O(100) times. It is a tremendous performance improvement, as visualized in Figure 5.1.

5.4.2 Mechanistic-Level DKM

Table 5.3 summarizes the model C8-HDC-Mech solving performance for various methods. LSODE and LSODA, when using a finite difference approach to evaluate the full Jacobian matrix internally, performed poorer than DASSL on this problem. However, by supplying the Jacobian explicitly, LSODE does better than DASSL in this case. In particular, LSODES, which utilizes the sparseness of the Jacobian matrix, does a much better job than all the other methods. (In model C8-HDC-M, the sparsity of its Jacobian matrix is only 8.5% [see Equation 5.9].) This tells us the significance of utilizing Jacobian sparseness in improving model solving performance. When the user supplies the sparse Jacobian explicitly, LSODES performs even better. The results also show that the Adams method does not even converge after an excessive amount of work (500 steps) is done, which indicates that the mechanistic level models are more stiff in nature. This makes sense because in mechanistic level modeling, both observable molecules and unobservable transient intermediates are represented in the reaction network, and there are big differences among the elementary-step rate constants for different reaction families, which leads to a very stiff problem.

For the mechanistic-level DKM in this experiment, the solver performance ranking is

LSODES with BDF and user-supplied Jacobian
> LSODES with BDF and finite-differenced Jacobian
>> LSODE with BDF and user-supplied Jacobian

> DASSL
> LSODE with BDF and finite-differenced Jacobian
> LSODA

Exploiting the compiler optimization over the debugging mode can further improve the model solving performance. Based on this work on mechanistic level DKMs, the optimized LSODES with the BDF method and user-supplied sparse Jacobian can get a performance gain over classical DASSL O(10) times. It is a significant performance improvement, as visualized in Figure 5.2. In the model optimization stage, when the model is solved thousands of times, the benefit of this performance gain becomes clearer. It can enable modelers to test ideas faster and be more creative.

5.4.3 DKM MODEL SOLVING GUIDELINES

As we have learned from the experiments, it is worthwhile and beneficial to do experiments with different algorithms, especially for DKM systems. Here are the three key model-solving guidelines for DKM systems.

1. First, mathematically identify and formulate the ODE system or DAE system; the ODE system tends to be solved faster with an ODE solver, rather than with a DAE solver.
2. Exploit the stiffness nature of the specific system — stiff algorithms such as BDF for stiff problems and nonstiff algorithms such as Adams–Moulton for nonstiff problems. Generally, pathways-level models are nonstiff in nature, so an Adams–Moulton algorithm would

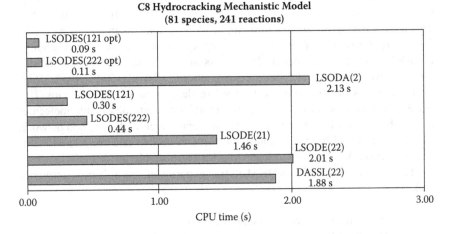

FIGURE 5.2 C8-HDC-Mech model solving performance for various solvers.

be preferred. Mechanistic-level models are stiff in nature, so a BDF algorithm would be preferred.

3. Exploit the sparseness of the specific system; taking advantage of the sparsity of a large reaction network can significantly improve the model solving performance.

5.5 SUMMARY AND CONCLUSIONS

In this chapter, we first reviewed the nature of the detailed kinetic models and the mathematical background to solve these systems. The underlying numerical methods, the stiffness of the DKM systems, and the sparseness of the DKM systems were reviewed.

Two candidate DKMs, one (GO-HDC-Path) for a pathway-level model and the other (C8-HDC-Mech) for the mechanistic-level model, and four candidate DAE and ODE solvers (DASSL, LSODE, LSODES, LSODA) were set up to test the model solving performance at various scenarios. All solvers were exploited to understand the underlying numerical methods to solve stiff and nonstiff ODE and DAE systems. The sparsity structure of the Jacobian matrix was exploited to improve the performance and accuracy of the numerical solutions. In our experiments, for the pathways-level DKM, the solver performance was ordered as LSODES with Adams > LSODE with Adams > LSODA >> LSODES with BDF >> LSODE with BDF >> DASSL; for the mechanistic-level DKM in this experiment, the solver performance was ordered as LSODES with BDF and user-supplied Jacobian > LSODES with BDF and finite-differenced Jacobian >> LSODE with BDF and user-supplied Jacobian > DASSL > LSODE with BDF and finite-differenced Jacobian > LSODA.

This work has demonstrated that it is very important to select appropriate numerical solvers for large DKMs, to exploit the stiffness of the specific physical problem, and to take advantage of the sparsity of large reaction networks. All of these can make significant differences in model solving performance. Generally, pathways-level DKMs are nonstiff, and the Adams–Moulton algorithm is preferred; mechanistic level DKMs are stiff, and the BDF algorithm is preferred. DKMs are generally sparse in nature, and therefore the solver that treats the Jacobian matrix's sparsity structure, such as LSODES, is recommended.

REFERENCES

Beris, A.N., Notes for *Chemical Engineering Problems*, Class notes, University of Delaware, Newark, 1998.

Brenan, K.E., Campbell, S.L., and Petzold, L.R., *Numerical Solution of Initial-Value Problems in Differential-Algebraic Equations*, North-Holland, Amsterdam, 1989.

Broadbelt, L.J., Stark, S.M., and Klein, M.T., Computer generated pyrolysis modeling: on-the-fly generation of species, reactions and rates, *Ind. Eng. Chem. Res.,* 33, 790–799, 1994.

Dente, M., Ranzi, E., and Goossens, A.G., Detailed prediction of olefin yields from hydrocarbon pyrolysis through a fundamental simulation model (SPYRO), *Comput. Chem. Eng.*, 3, 61–75, 1979.

Hindmarsh, A.C., et al., ODEPACK documentation, Lawrence Livermore National Laboratory, Livermore, CA, 1997.

Hindmarsh, A.C., personal communication, 1998.

Hou, G. and Klein, M.T., Automated Molecule-Based Kinetic Modeling of Complex Processes: A Hydroprocessing Application. Paper presented at 2nd International Conference on Refinery Processes Proceedings, AIChE, Houston, TX, 1999.

Joshi, P.V., Hou, G., and Klein, M.T., Molecule-Based Reaction Engineering: Interfacing Fundamental and Process Chemistry. Paper presented at 1st International Conference on Refinery Processes Proceedings, AIChE, New Orleans, LA, 1998.

Quann, R.J. and Jaffe, S.B., Building useful models of complex reaction systems in petroleum refining, *Chem. Eng. Sci.*, 51(10), 1615, 1996.

Quann, R.J. and Jaffe, S.B., Structure-oriented lumping: describing the chemistry of complex hydrocarbon mixtures, *Ind. Eng. Chem. Res.*, 31(11), 2483, 1992.

Radhakrishnan, K. and Hindmarsh, A.C., Description and Use of LSODE, the Livermore Solver for Ordinary Differential Equations, NASA Reference Publication, Lewis Research Center, Clever and OH, 1327, 1993.

Sundaram, K.M. and Froment, G.F., Modeling of thermal cracking kinetics. I, *Chem. Eng. Sci.*, 32, 601–608, 1977a.

Sundaram, K.M. and Froment, G.F., Modeling of thermal cracking kinetics. II, *Chem. Eng. Sci.*, 609–617, 1977b.

6 Integration of Detailed Kinetic Modeling Tools and Model Delivery Technology

6.1 INTRODUCTION

The discipline of chemical engineering provides a rigorous framework for the construction, solution, and optimization of detailed kinetic models for delivery to process chemists and engineers. In Chapter 2, we discussed the molecular structure and composition modeling techniques used to transform analytical chemistry information into molecular representations of complex feedstocks through stochastically sampling molecular attributes. In Chapter 3, we discussed how to automatically build and control the reaction network for complex process chemistries on the computer via various algorithms and strategies. In Chapter 4, we discussed how to organize and evaluate thousands of rate parameters in the complex reaction network from fundamentals-based chemical engineering kinetics correlations. In Chapter 5, we discussed the model solution techniques to solve the detailed kinetic models fast and accurately. The goal of this chapter is to integrate all the above technical components of detailed kinetic modeling, namely, the modeling of reactant structures and composition, the building of reaction networks, the organization of model parameters, the solution of the kinetic model, and the optimization of the model (this chapter), which are often in a vague and incomplete form requiring assumptions and approximations for use. This integrated chemical engineering software package is named the Kinetic Modeler's Toolbox (KMT).

An overlying issue, from the chemical engineering software development point of view, is that the model delivery must be in a form that makes the model accessible to process chemists and engineers who may not be experts in computer hardware, operating systems, and programming languages. We will thus address KMT development and use issues in the context of model delivery to serve the end users' needs. We will also document the issues and lessons learned in the development process of this complex software package.

This chapter is thus divided into two parts. The first part discusses, more from the chemical engineering perspective, the integration of the technical components of detailed kinetic modeling into a single user-friendly software package, KMT, and the optimization framework on which KMT sits. The second part is more focused on the software development side of the work, which documents the development and use issues on how to make the KMT accessible and easy-to-use on routine platforms for process chemists and engineers.

6.2 INTEGRATION OF DETAILED KINETIC MODELING TOOLS

In Chapter 1, we introduced a modeling strategy to solve the complex process modeling problem via a different route at the fundamental molecular level. In modeling the petroleum refining process, we would like to be able to predict the product properties from the feedstock characterization, that is, feedstock characterization as the input and product properties as the output.

Figure 6.1 shows the integrated molecule-based kinetic modeling approach in KMT. In this section, first we will discuss the KMT integration module by module and the interconnection between these modules. Then, we will discuss the optimization framework and property estimation that KMT uses to match the model results with experimental observations.

6.2.1 THE INTEGRATED KINETIC MODELER'S TOOLBOX

As shown in Figure 6.1, KMT has five modules to automate the kinetic modeling process: the molecule generator (MolGen), the reaction network generator (Net-Gen), the model equation generator (EqnGen), the model solution generator (SolGen), and the parameter optimization framework (ParOpt). It begins with molecular structure building software (MolGen) that uses Monte Carlo simulation techniques to assemble a molecular representation of complex feedstocks from analytical information (e.g., the hydrogen-to-carbon ratio [H/C], SIMDIS, NMR, etc.). Then, graph theory techniques are utilized to generate the reaction network (NetGen). Reaction family concepts and quantitative structure reactivity correlations (QSRCs) are used to organize and estimate rate constants. The computer-generated reaction network, with associated rate expressions, is then converted to a set of mathematical equations (EqnGen), which can be solved for different reactor systems (SolGen) within an optimization framework to determine the rate parameters (ParOpt) in the model. This automated molecule-based kinetic modeling process enables the modeler to focus on the fundamental chemistry and significantly speed up the model development.

Each module will be discussed in the following subsections. The parameter optimization will be discussed in more detail in Section 6.2.2.

6.2.1.1 The Molecule Generator (MolGen)

The molecule generator (MolGen) is basically the realization of the stochastic modeling techniques for molecular structure and compositions of complex feedstocks, as discussed in Chapter 2. From a modeling point of view, there are two kinds of inputs to MolGen: adjustable and fixed. The analytical information such

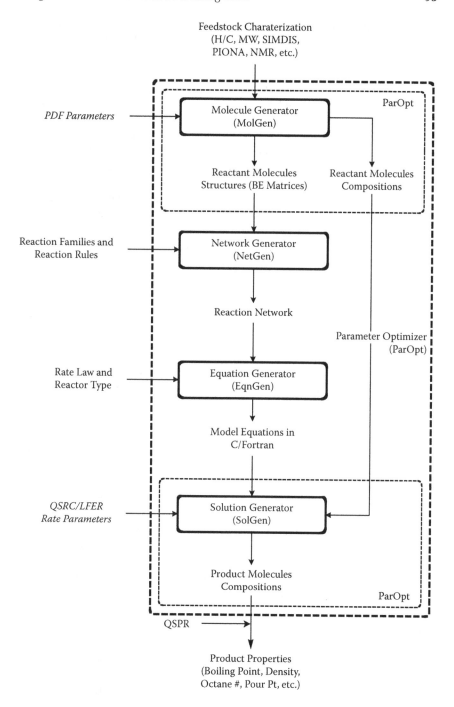

FIGURE 6.1 An integrated detailed Kinetic Modeling Toolbox (KMT).

as the hydrogen-to-carbon (H/C) ratio, average molecular weight, boiling point distribution, compound class distribution, and NMR, are fixed characterizations for a fixed feedstock. The adjustable characteristics are parameters of the possibility distribution functions (PDFs) used to stochastically sample the molecular attributes, such as number of aromatic rings, the number of naphthenic rings, the length of side chains, the number of side chains, etc. MolGen can be solved once with fixed PDFs, but it is normally solved within an optimization framework (ParOpt), as depicted in Figure 6.1, to adjust and search the optimal PDF parameters to match the analytical observations for the feedstock. The feedstock characterizations can be readily calculated or estimated from the explicit molecular structures and compositions via the quantitative structure property relationships. The outputs from MolGen include the representative molecular structures and their optimized compositions for the feedstock. The reactant molecular structures are then converted to the corresponding bond–electron (BE) matrices in the grammar of the NetGen and serve as its input. The composition or concentrations of these reactant molecules will be the initial conditions for the quantitative model later in the SolGen stage.

6.2.1.2 The Reaction Network Generator (NetGen)

The first version of the reaction network generator (NetGen), developed by Broadbelt et al. (1994a, 1996), was for ethane pyrolysis (Broadbelt et al., 1994b). The current version of the NetGen module in KMT is an enhanced generation with additional features and functionalities as discussed in Chapter 3. NetGen has been enhanced to handle more complex molecular structures, such as the multiring polynuclear aromatic, hydroaromatic, sulfur, and nitrogen compounds, which we will demonstrate in the gas oil hydroprocessing application (Chapter 10). In terms of chemistry, it has been enhanced to handle much more complex multifunctional heterogeneous chemistries, such as the metal chemistry and acid chemistry at both the pathways and mechanistic level, which we will demonstrate in the newly developed applications including naphtha reforming (Chapter 7), heavy paraffin hydrocracking (Chapter 8), naphtha hydrotreating (Chapter 9), gas oil hydroprocessing (Chapter 10), and gas oil fluid catalytic cracking (FCC) (Chapter 11), as well as naphtha pyrolysis (Chapter 12) in Part II.

The fundamentals are based on graph theory. Molecules and intermediates (ions and radicals) are represented as BE matrices. Reactions are carried out via simple matrix addition operations between the reactant matrix and the reaction matrix for each reaction family to generate the product matrix. Details can be found in Chapter 3.

There are two categories of inputs to NetGen. One category is the reaction molecules in the format of BE matrices. For a simple model compound, its BE matrix can be easily written by following the grammar. For complex feedstocks, the representative molecular structures can be generated from MolGen, as depicted in Figure 6.1. The other category of inputs to NetGen is the reaction chemistry including both the reaction families in the chemistry and reaction rules to control the model size based on the user's experience and understanding. NetGen takes in both inputs and applies the reaction matrix for each reaction family on top of each reactant molecule and the reaction rules on top of each reaction to build up the complete reaction network. The details of the automated model building algorithms can be found in Chapter 3. The generated reaction network is written into a file in the required format of EqnGen and serves as its input.

6.2.1.3 The Model Equation Generator (EqnGen)

The model equation generator (EqnGen) originates from the previous OdeGen (Broadbelt et al., 1994b), which converts the reactions into the corresponding ordinary differential equations. EqnGen extends the capabilities to generate different kinds of mathematical equations from the reaction network based on the rate law and the reactor types. For the batch reactor (BR), the plug flow reactor (PFR), or the fixed bed reactor, EqnGen writes out the corresponding ordinary differential equations, or differential algebraic equations for situations with steady state assumptions. For the continuous stirred tank reactor (CSTR), it writes out the corresponding algebraic equation systems. For different rate laws, such as the Langmuir–Hinshelwood–Hougen–Watson (LHHW) formalism, EqnGen can write different rate constant functions. Essentially, EqnGen is a mathematical converter that parses the reaction network and generates the corresponding mass balance equations for each species in the system. The system of generated mass balance equations forms the kernel of the kinetic model in the context of a reactor model. The code for the model equations can be generated either in C or FORTRAN.

6.2.1.4 The Model Solution Generator (SolGen)

The model solution generator (SolGen) is the combination of equations generated from EqnGen, coupled with a process driver that is embedded with some equation system solver, such as the Livermore Solvers for Ordinary Differential Equations (LSODE) and its variants (Radhakrishnan and Hindmarsh, 1993; Hindmarsh et al., 1997) for solving ordinary differential equations (ODEs) or *d*ifferential/*a*lgebraic *s*ystem *solv*er (DASSL) (Petzold, 1982; Brenan, et al., 1989) for solving differential algebraic equations (DAEs), as discussed in Chapter 5. The driver file sets up all the I/O files and reactor configurations and calls the model equations, which can be compiled and run to produce the final solution. The driver file is specific for each process because of the slight differences between different process configurations. For example, a 3-PFR process configuration used in naphtha reforming (Chapter 7) needs a slight change in the driver file for PFR to accommodate this configuration. The process driver file, the model equations file, the file for supporting functions, and various I/O files form the model deliverable that can be compiled and solved together to produce the model results.

The initial conditions of the model, such as the concentrations of each reactant molecule, can come directly from the experimental measurement or from the optimized results from MolGen for complex feedstocks. Generally, the only adjustable inputs to the model are those rate parameters, that is, the quantitative structure reactivity correlations (QSRC) or linear free energy relationship (LFER) parameters for each reaction family, as discussed in Chapter 4. The model can be solved in once-through mode with supplied rate parameters to generate the concentration profiles of all product molecules along the reaction time for a BR or along the reaction length for a PFR. However, the rate parameters are not always known *a priori* for many process chemistries, especially the heterogeneous catalytic processes studied in this work. Therefore, the model can also be solved in tuning mode. In this mode, it is alternately solved and tuned within an optimization framework to match the experimental observations by adjusting and optimizing the rate parameters. The parameter optimization framework (ParOpt) and the strategy for matching the model results with the observable or interested product properties will be discussed in Section 6.2.2. Finally, the tuned kinetic model can be used to predict various product properties at various process operation conditions.

6.2.2 PARAMETER OPTIMIZATION AND PROPERTY ESTIMATION

Parameter optimization and property estimation techniques are integral parts of the KMT. In this section, we will briefly review the parameter optimization algorithms used in KMT and the issues related to the parameter optimization and property estimation.

6.2.2.1 The Parameter Optimization (ParOpt) Framework

The Monte Carlo simulation at the MolGen module is normally run within an optimization framework by adjusting the PDF parameters to allow the stochastically generated molecular structures to match both the structures and compositions with the feedstock. The developed kinetic model template can be solved within an optimization framework to determine the kinetic rate parameters by matching the model results with the experimental observations in the reactor. Both of the above parameter optimization (ParOpt) cases have been noted with dashed rectangles in Figure 6.1 as part of the KMT.

6.2.2.2 Optimization Algorithms

Many optimization algorithms have been tested with and integrated into the optimization framework of KMT. The three major optimization algorithms we have frequently used with KMT are the simulated annealing (SA) method, the GREG routine, and the multilevel single linkage method.

The SA method, since its inception in the early 1980s (Kirkpatrick et al., 1983), has become a powerful tool for global optimization of large-scale systems. Much work has documented the theory and applications of SA, such as that by Laarhoven et al. (1987). SA is rooted in thermodynamics. The process of heating and slow cooling (annealing) of a solid to remove strain and crystal imperfections motivates this algorithm. During this process the free energy of the solid is minimized. The initial heating is necessary to avoid converging a local minimum. Every function can be viewed as the free energy of some system, and the SA method imitates the way nature reaches a minimum during the annealing process.

KMT, or more specifically, the MolGen module, has implemented the SA algorithm for continuous variable problems developed by Corana et al. (1987) and implemented by Goffe et al. (1991). MolGen uses SA to find the optimal O(10) PDF parameters by minimizing the objective function that matches the representative molecular structures and compositions with the feedstock characteristics. Due to its global optimization, SA does not converge very fast, but it is very accurate: a typical MolGen optimization normally takes hours to days on an IBM 560 server (66-MHz CPU). This meets the requirements of MolGen, where we need an accurate molecular representation and time is not an issue since this optimization is normally a one-time, offline effort in a model development process. We only need to run it once to get the reactant molecular structures and compositions for further model development. After a library of analyzed feedstocks is built, the optimization of a slightly changed feed would be much faster.

GREG, developed by Sørensen (1982), Caracotsios (1986), and Stewart et al. (1992a), is a nonlinear parameter estimator that uses a Bayesian approach to estimate model parameters and their inference intervals and covariances, using single-response or multiresponse data. Multiresponse data are familiar outputs of experiments and processes involving multicomponent mixtures, as in our detailed kinetic modeling. Applications of GREG for chemical engineering problems, including kinetic models, can be found in the literature, e.g., Stewart et al. (1992a). Due to its local optimization, GREG is much faster than SA. This is the main reason GREG was chosen to optimize the rate parameters in the kinetic model. The model needs to be solved repetitively in every iteration of the optimization process, so the speed of the optimization is crucial. A typical round of SolGen optimization can normally take from 0.5 hour for a pathways-level model to many hours for a mechanistic model on a 400-MHz Pentium II PC.

The multilevel single linkage (MLSL) algorithm, developed by Rinnooy Kan and Timmer (1984a), is a global optimization procedure that uses stochastic sampling iteratively to acquire information about an objective function surface. It is a versatile optimization method because it is able to use virtually any local minimization procedure (both direct and gradients methods) to locate the objective function local minimum. This method has been shown to have excellent convergence and efficiency properties. The fundamental idea behind the MLSL algorithm is that an objective function is composed of a collection of local minima, each of which has associated regions of attraction. A minimum's region of attraction is defined by the set of parameter vectors that are mapped to a local minimum by a minimization procedure. Detailed information can be found in Chapter 3 of Stark (1992) and references within. MLSL, as implemented by Stark (1992, 1993), was used in KMT, both in MolGen and SolGen. For relatively large kinetic models, the gradient method is preferred to search for local minima because of its better performance compared with direct methods. In the development of detailed kinetic models, the MLSL global optimizer was used in parallel with other optimization routines to generate the initial guess values and the parameter ranges for other local parameter optimizers in order to speed up the overall optimization process.

6.2.2.3 The Objective Function

In all the above-mentioned optimization routines, one important element that users can adjust is the objective function. This is crucial for the successful optimization and best results of the optimized parameters.

In Chapter 2, we discussed the objective function used to optimize the molecular representation with the actual feedstock characterizations: the chi-square statistic (χ^2) as defined in Equation 2.4. The numerator is the square of the difference between the model prediction (i.e., MolGen-generated molecules properties) and the experimentally determined properties. The denominator is a weighting factor equal to the standard deviation of the experimentally determined value. This objective function can be modified easily for any analytical information on a feed. The more analytical information we have on the feedstock, the better the objective function will be. Lower values of the objective function indicate that a molecular representation matches the experimental data better.

To optimize the rate parameters for a developed kinetic model, the objective function is normally defined as the square of the difference between predicted and experimental yields weighted by the experimental standard deviation as shown in Equation 6.1

$$F = \sum_{i=1}^{M} \sum_{j=1}^{N} \left(\frac{y_{ij}^{model} - y_{ij}^{exp}}{\bar{\omega}_j} \right)^2 \qquad (6.1)$$

where i is the experiment number, j is the species or lump number, and ωj is the weighting factor (generally the experimental measurement deviation). The assignment of ωj is very important for the success of the optimization. Generally, the ωj should be the same magnitude as y or less to make sure F/MN $\leq$ 1. The smaller the ωj, the more important it would be in the objective function. Sometimes, the choice of ωj can trade off with the user's expectation. For example, as in hydrotreating or hydrodesulfurization (HDS) process (Chapter 9), if the user cares more about the sulfur content in the products than some paraffins, then the weighting factor ωj should be set to very small for sulfur to better predict the sulfur content.

6.2.2.4 Property Estimation of Mixtures

The direct solution from a kinetic model is normally the concentrations of all the product molecules. However, in petroleum refining processes, the process chemists and engineers are more interested in mixture properties that characterize the combustion quality, fluid characteristics, or polluting potential, etc., such as the octane number, cetane number, pour point, smoke point, freeze point, cloud point, diesel index, refractive index, viscosity index, and sulfur and nitrogen content. The mixture property could be calculated based on the collective properties of the product molecules as long as the molecular properties are available in addition to the molecular compositions predicted from our model.

The strategy is thus to develop or utilize any quantitative structure property relationships (QSPRs) to correlate the molecular structure with the molecular property. Generally, the molecular properties exhibit a strong correlation with carbon number for each homologous series of molecular class. There exist many empirical and well-tested correlations in the literature to calculate many of these properties, such as the Joback group contributions method to estimate boiling point from the core ring structure and additional CH_2- groups. The QSPRs are only needed for the essential properties, specifically the boiling point and density (Quann and Jaffe, 1996). Their mixture properties can then be calculated from the stream compositions, and many other mixture properties can generally be estimated from them using a wealth of correlations in the literature. Zhang and Towler (1999) have demonstrated examples of the molecular structure–property correlations to predict both density and boiling point. The ultimate source is the Beilstein Institute for Literature of Organic Chemicals, which compiles all the property data for pure components containing those molecular groups in the petroleum process streams.

6.2.2.5 The End-to-End Optimization Strategy

We have discussed both the optimization of PDF parameters at the MolGen stage and the optimization of kinetic (QSRC/LFER) parameters at the SolGen stage. Furthermore, in principle, it is clear from Figure 6.1 that we can optimize all the parameters together since we have automated the whole kinetic model development process from the beginning of feedstock characterization to the end of product properties. As long as both the reaction chemistry (input to NetGen) and the reaction kinetics (input to EqnGen) are fixed, we can optimize the PDF parameters and kinetic parameters together to predict the product properties from feedstock characterizations. This end-to-end optimization strategy would be feasible with the evolution of KMT and our increasingly better understanding of the reaction chemistry and reaction kinetics along the way.

The optimization framework to tune the parameters is actually necessary to build useful models. This is because the kinetics determined with model compounds in the lab may not be representative of commercial operations due to the catalyst and reactor factors. Even the model derived from the fundamentals must have the capability of tuning kinetic parameters to accommodate any irregularities of the specific process and match the commercial performance.

6.2.3 Conclusions

All the detailed kinetic modeling tools are integrated into one software package, the Kinetic Modeler's Toolbox (KMT), to automate the modeling of molecular structures and kinetics of complex reaction systems. KMT includes a total of five modules: MolGen for molecular structure and composition modeling, NetGen for automated reaction network generation, EqnGen for automated code and equation generation, SolGen for the model solution, and the parameter optimization framework, ParOpt. This automated molecule-based modeling approach delineates and reduces the essential elements of the complexity in the complex reaction systems to a manageable and irreducible level: the molecular structures are reduced to $O(10)$ PDF parameters (five to ten PDFs times two to three parameters in each PDF), the reaction network building is automated through $O(10)$ reaction matrices (one for each reaction family), and the molecular reactivities are correlated with $O(10)$ kinetic parameters (five to ten LFER's times two to three parameters each). The brute force description of these complex systems could have been of the order of $O(10^5 \times 10)$ due to the $O(10^5)$ species and $O(10)$ reactions from each species in these complex reaction systems, which can easily become out of control.

The integrated KMT in Figure 6.1 gives a clear picture of the kinetic modeling approach from the input of feedstock characterization to the estimation of output product properties. The set of the PDF parameters is like a fingerprint of the complex feedstock and can be optimized to assemble its molecular representation. Our understanding of the reaction chemistry provides us the reaction families and reaction rules, and our understanding of the reaction kinetics provides us the rate law. The set of QSRC/LFER parameters fundamentally correlates the molecular reactivities with the molecular structures.

The KMT software has automated the process of building these detailed kinetic models. By exploiting Monte Carlo and graph theory techniques, reaction models containing thousands of species can be built in 1000 CPU seconds or less. This model building speed has changed the serial model building–model using paradigm to a new parallel approach, where a model builder can produce an updated optimal model in seconds. These models can then react to the molecularly explicit feedstock using QSRCs for kinetic parameters to predict the product properties.

6.3 KMT DEVELOPMENT AND MODEL DELIVERY

The KMT can be used for either R&D or commercial process optimization. In either case, it is important that the toolbox and the developed models be easily usable. In this section, we will discuss, from the software development perspective, various issues related to KMT development and use to make the tools and models accessible and easy to use on routine platforms and to make the information more useful and insightful for process chemists and engineers.

6.3.1 PLATFORM AND PORTING

The early form of KMT or, more specifically, the first version of the reaction network generator (NetGen), developed by Broadbelt et al. (1994a, 1996), ran on an IBM RISC 6000 server with the AIX operating system (an IBM version of UNIX). One round of model building, solving, and optimization process could take a couple of days to a couple of weeks to complete. This significantly hindered the model development process since to test even one idea could take weeks. Also, NetGen was accessible to experts only, rather than their targeted users: process chemists and engineers who may not be experts in computer hardware and operating systems. This motivated us to develop KMT in a more accessible mode.

The KMT delivery platform of PC/Windows was chosen for many reasons. Most process chemists and engineers are more familiar with a PC/Windows environment. A low-end PC would have a much higher performance-to-cost ratio than even a high-end server or mainframe servers. Windows is easy to use because of its graphic user interface. The current version of KMT ported to PC/Windows platform works well. On a Pentium Pro 200-MHz machine, the detailed kinetic models at the pathways level can usually be solved in seconds; mechanistic models can be solved in minutes. The most time-consuming parameter optimization takes only hours to complete. This has significantly accelerated the model development process at a low cost and in a highly efficient manner.

The porting of KMT was a nontrivial job. The whole package of KMT includes hundreds of files written in a mixed programming environment with C, C++, and Fortran over a period of years. An innovative porting path — from IBM server/AIX to PC/Linux then to PC/Windows as shown in Figure 6.2 — was chosen to make this happen. All the code developed under AIX was modified and compiled under the GNU C/C++/Fortran compilers, which are normally shipped with most Linux systems (open source versions are available from the Free Software Foundation at http://www.gnu.org/) rather than the proprietary IBM compilers on AIX. The code written under AIX was then ported to a Linux operating system and compiled with GNU compilers. Since both AIX and Linux are UNIX variants, the code change and porting complexity were manageable. The major change in the code is the machine constants for the model solvers resulting from the change from IBM RISC 6000 architecture to PC Intel architecture. The next step was to port from Linux to Windows with both running on a PC. This step was achieved with the assistance of CygWin (formerly Cygnus). (A complete set of free GNU UNIX-Work Alike tools compiled for Windows can be downloaded from the Internet at http://www.cygnus.com/.) The current version of KMT and the final kinetic models can run under all three major

FIGURE 6.2 The server/UNIX platform to the PC/Windows platform porting approach.

platforms, that is, server (both IBM and SUN)/UNIX (both AIX and Solaris), PC/Linux, and PC/Windows (Windows 2000, NT, and 9x).

With today's high CPU performance of PCs and the significant improvement of algorithms (Chapter 5), both the model building and model solving time are so fast as not to be a limitation anymore. Many candidate models can thus be built to test all kinds of ideas and assumptions. The insights gained by testing the mechanistic models can be used to simplify the reaction network at the pathways level, as will be shown in the heavy paraffin hydrocracking application in Chapter 8. This iterative model development process is thus streamlined, and the detailed kinetic models can be updated in seconds as user knowledge becomes available.

6.3.2 DATA ISSUES

When viewed from the perspective of modeling, most data are not in a usable form. There are more numbers than useful information, and the useful information is sparse. A significant effort is needed to clean and organize the data for kinetic modeling. During the automated kinetic modeling process, the most time-consuming step is the data organization for model tuning. The model building only takes seconds, whereas the data organization takes weeks or months.

The current version of delivered kinetic models uses the following I/O files to organize the data. There are three required input files: the rate parameters file (.rtk) for all the adjustable kinetic parameters, the model input files (.inp) for the reactor information and initial reactant concentrations, and the molecular information file (.inf) generated from NetGen for all molecular structural (such as C#, H#, MW, etc.) and organizational information (such as lump ID). All these files can be modified as needed. There are two files for model optimization: the experimental observations file (.obs) for model comparison and the weighting factors file (.wgt) for the weights of observation lumps in the objective function. There is also an optional file for recycle composition (.rcy). Two files are used for organizing the output: one is for the user-interested lumped output only (.out), and the other is for all the molecular and intermediate compositions along the trajectory (.ou1). All the raw data files are in text format that can be imported directly into any spreadsheet program such as Excel for further processing. The standard functions such as the macro and filtering function in Excel provide enough intellectual flexibility for a normal user to analyze the data and generate useful graphs for R&D purposes. These data files include all the basic information of what a molecule-based kinetic model needs and outputs. A further refinement can be made to formalize the standard data gathering and representation template for detailed kinetic models.

6.3.3 USER INTERFACE ISSUES

Two kinds of issues are related to the user interface (UI): one is the standard look and feel, and the other is the intellectual organization of I/O at the front end. The former is a software development issue, and the latter is a chemical engineering, intellectual issue. Regarding the look and feel, two kinds of efforts have been pursued: the use of Microsoft Excel and the use of a standard Web browser, such as Internet Explorer or Netscape Navigator. At the front end (either using Excel or a Web browser), the inputs are collected and written into the designated text files while the model is running in the background. Then Excel or a CGI/BIN program for a Web server, such as Perl, reads from the model output files and makes a display at the front end. Regarding the intellectual organization, the layout and I/O should be familiar to the users and designed for their needs, and the basic I/O should be apparent. If the model is targeted to process chemists and engineers, the process model user interface should have the same configuration and operation as the commercial unit and simulate the process. The units of measure should be based on the conventional characterization methods available in the refinery.

The integrated KMT provides much more flexibility in terms of what a user can do since the user can explore all the inputs shown in Figure 6.1. At the background of KMT, "on-the-fly" compiling of the code and automation scripts hides the model development process.

6.3.4 DOCUMENTATION ISSUES

The automated model building feature of KMT makes documentation rather straightforward. MolGen can assemble the representative molecules and generate their bond–electron matrices. NetGen can produce a complete list of species with their structural information and a complete reaction network. These species and reactions can be easily organized into molecular classes and reaction families. SolGen, for example, for a PFR case, can produce the trajectory of molecular concentrations along the reactor, which can be used to generate the graphs such as y~t, s~x on-the-fly via gnuplot on Linux/UNIX or Excel on Windows. All these documentations can be easily converted to HTML format and presented on the Web. The documentation is thus merged with the front end on the Web.

6.3.5 LESSONS LEARNED

The KMT development process is a reasonable example of interdisciplinary research conducted by marrying the fundamental research and pragmatic attitude of chemical engineering with the rigorous software development process of computer science. The development of such a large software system requires both the depth of knowledge in molecular reaction engineering and the rigorous software engineering practice. Many lessons have been learned throughout this process, such as the importance of version control and software maintenance (Bennett et al., 1990).

6.4 SUMMARY

This chapter has integrated all the detailed kinetic modeling tools together into one user-friendly software package — the Kinetic Modeler's Toolbox (KMT). KMT has five modules used to automate the molecule-based kinetic modeling process: the molecule generator (MolGen), the reaction network generator (Net-Gen), the model equation generator (EqnGen), the model solution generator (SolGen), and the parameter optimization framework (ParOpt). This automated kinetic modeling process enables the modeler to focus on the fundamental chemistry and significantly speed up the model development.

REFERENCES

Bennett, K.H., The software maintenance of large software systems: management, methods and tools, in *Software Engineering for Large Software Systems,* Kitchenham, B.A., Eds., Elsevier, New York, 1990.

Brenan, K.E., Campbell, S.L., and Petzold, L.R., *Numerical Solution of Initial-Value Problems in Differential-Algebraic Equations,* North-Holland, Amsterdam, 1989.

Broadbelt, L.J., Stark, S.M., and Klein, M.T., Computer generated reaction networks: on-the-fly calculation of species properties using computational quantum chemistry, *Chem. Eng. Sci.*, 49, 4991–5101, 1994a.

Broadbelt, L.J., Stark, S.M., and Klein, M.T., Computer generated pyrolysis modeling: on-the-fly generation of species, reactions and rates, *Ind. Eng. Chem. Res.,* 33, 790–799, 1994b.

Broadbelt, L.J., Stark, S.M., and Klein, M.T., Computer generated reaction modeling: decomposition and encoding algorithms for determining species uniqueness, *Comput. Chem. Eng.*, 20(2), 113–129, 1996.

Caracotsios, M., Model Parametric Sensitivity Analysis and Nonlinear Parameter Estimation: Theory and Applications, Ph.D. thesis, University of Wisconsin, Madison, 1986.

Corana, A., Marchesi, M., Martinii, C., and Ridella, S., Minimizing multimodal functions of continuous variables with the "simulated annealing" algorithm, *ACM Trans. Math. Software,* 13(3), 262–280, 1987.

Edelson, D. and Allara, D.L., Parameterization of complex reaction systems: model fitting vs. fundamental kinetics, *AIChE J.,* 19(3), 638, 1973.

Goffe, W.L, Ferrier, G.D., and Rogers, J., Global optimization of statistical functions with the simulated annealing algorithm, *J. Econometrics,* 60, 12, Jan-Feb, 65-100, 1994.

Hindmarsh, A.C., et al., ODEPACK Documentation, Lawrence Livermore National Laboratory, Livermore CA, 1997.

Kirkpatrick, S., Gelatt, C.D. Jr., and Vecchi, M.P., Optimization by simulated annealing, *Science,* 220(4598), 671, 1982.

Petzold, L.R., A description of DASSL: a differential/algebraic system solver, *Sandia Tech. Rep.,* SAND82-8637, 1982.

Quann, R.J. and Jaffe, S.B., Building useful models of complex reaction systems in petroleum refining, *Chem. Eng. Sci.,* 51, 1615–1635, 1996.

Radhakrishnan, K., Hindmarsh, A.C., Description and Use of LSODE, the Livermore Solver for Ordinary Differential Equations, NASA Reference Publication 1327, 1993.

Rinnooy Kan, A.H.G. and Timmer, G.T., Stochastic methods for global optimization. I. Clustering methods, *Math. Programming,* 39, 27–56, 1984a.

Rinnooy Kan, A.H.G. and Timmer, G.T., Stochastic methods for global optimization. II. Multi level methods, *Math. Programming,* 39, 57–78, 1984b.

Sørensen, J.P., Simulation, Regression and Control of Chemical Reactors by Collocation Techniques, doctoral thesis, Danmarks tekniske Højskole, Lyngby, 1982.

Stark, S.M., An Investigation of the Applicability of Parallel Computation to Demanding Chemical Engineering Problems, PhD dissertation, University of Delaware, Newark, 1992.

Stark, S.M., The Definitive Guide to the MLSL Program, Dept. of Chemical Engineering, University of Delaware, 1993.

Stewart, W.E., Caracotsios, M., and Sørensen, P., GREG Software Package Documentation, Dept. of Chemical Engineering, University of Wisconsin, Madison, 1992a.

Stewart, W.E., Caracotsios, M., and Sørensen, J.P., Parameter estimation from multiresponse data, *AIChE J.,* 38(5), 641, 1992b.

van Laarhoven, P.J.M. and Aarts, E.H.L., *Simulated Annealing: Theory and Applications,* D. Reidel Publishing Company, Dordrecht, Holland, 1987.

Zhang, Y. and Towler, G.P., Characterization and Property Prediction of Petroleum Mixtures, paper presented at the AIChE Spring Meeting, Houston, 1999.

Part II

Applications

7 Molecule-Based Kinetic Modeling of Naphtha Reforming

7.1 INTRODUCTION

Part I of this book reviewed the integrated chemical engineering tools for the automated building, solution, and delivery of detailed kinetic models. Part II will cover the application of these tools to the development of detailed kinetic models for various industrial complex processes, including naphtha reforming (this chapter), paraffin hydrocracking (Chapter 8), naphtha hydrodesulfurization (Chapter 9), gas oil hydroprocessing (Chapter 10), gas oil fluid catalytic cracking (Chapter 11), and naphtha pyrolysis (Chapter 12). The naphtha-reforming model building is a continuation of the C8 detailed naphtha modeling work of Joshi (1998).

Catalytic reforming is an important refinery process designed to boost the octane number of naphtha and also to produce aromatics such as benzene, toluene, and xylenes. The standard naphtha feedstock is a mixture of hydrocarbons normally in the range of C5 to C12 and can be as high as C14. A heterogeneous bifunctional catalyst (both metal and acid functions, such as Pt on alumina) is generally used for this process (Antos et al., 1995).

Various models have been developed (Kmak et al., 1971; Marin et al., 1982; Ramage et al., 1987; Mudt et al., 1995). In the 1990s, reforming processes were carefully scrutinized in terms of their molecular product composition, for example, for the total and molecular-level aromatic make (Antos et al., 1995). The arrival of new catalysts and thus new chemistries motivates the enhancement of these early modeling approaches. The new paradigm is to track each molecule in the feed and product through the process.

Developing a model with detailed kinetic information requires modeling of the chemistry at either the pathways level or the mechanistic level. Pathways models are not as complex as mechanistic models because of the exclusion of reaction intermediates, for example, carbenium ions, and offer the advantage of fast solution and optimization. This allows for their uses in online process optimizations and control. Pathways models offer the promise of containing detailed kinetic information by the explicit inclusion of observable molecules and hence, have the predictive capabilities of feedstock independence that lumped models lack. Various mechanistic insights into the chemistry can be incorporated into the reaction pathways too.

Pathways models can also be quite large and difficult to build with the increase of carbon number. On the one hand, the need for detailed molecular and kinetic information motivates the development of molecule-based explicit models; on the other hand, the size of these models is formidable to formulate manually. In this work, we will demonstrate how to apply the automated kinetic modeling process to the catalytic naphtha reforming process.

This chapter is presented in several sections. The next section describes the catalytic reforming chemistry at the reaction pathways level, and the modeling approach with the associated reaction matrices and reaction rules. The following section discusses the technical specifications and details for each reaction family. Then the model building process is automated, and a fine-tuned C14 catalytic naphtha reforming model is presented and validated. Finally, we will summarize the key observations and make some conclusions.

7.2 MODELING APPROACH

In the construction of detailed kinetic models, two key issues must be addressed. First, the chemistry must be represented in a rigorous manner that will allow the computer algorithm to handle the tedious job of keeping track of the species and reactions. Second, the rate parameters must be organized in order to represent the kinetics and obtain true rate constant information after optimization with the experimental data.

As we discussed in Chapter 2 and 3, much of the complexity is statistical or combinatorial, and the large demand for reactions and rate parameters in the molecule-based kinetic models can be handled by organizing the reactions involving similar transition states into one reaction family, the kinetics of which is constrained to follow the linear free energy relationship (LFER) or quantitative structure-reactivity correlations (QSRC).

Applying these ideas to naphtha reforming starts with an examination of its feedstock. The reforming reaction mixture can be grouped into a few compound classes (paraffins, isoparaffins, five- and six-membered naphthenes, and aromatics), which in turn react through a limited number of reaction families (dehydrocyclization, paraffin isomerization, hydrocracking, hydrogenolysis, naphthene isomerization, dehydrogenation, and dealkylation). As a result, a small number of reaction matrices can be used to generate the reaction network with hundreds of reactions, as discussed in Chapter 2.

The division of all the reforming reactions into a small number of reaction families with an associated reaction matrix can be automated in order to exploit the repetitive nature of the operations. Clearly, the use of the computer algorithm in the formulation of the model is very important, thus allowing researchers to focus on the basic chemistry and reaction rules of the model. In the next sections, graph-theoretic concepts are exploited to automate the reforming reaction network generation. The reactions pathways in the naphtha reforming process and the corresponding reaction matrices for all reaction families are summarized in Figure 7.1 (Joshi et al., 1999) and Table 7.1, respectively.

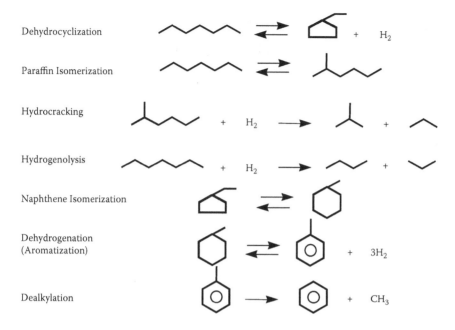

FIGURE 7.1 Naphtha reforming reaction families at the pathways level.

In this application of reforming kinetics at the pathways level, each sterically similar reaction family is defined by a single Arrhenius A factor, and the activation energy is correlated with the heat of reaction as shown in Equation 7.1.

$$E^* = E_o^* + \alpha \Delta H_{rxn} \qquad (7.1)$$

As discussed in Chapter 4, adsorption parameters can also be correlated in terms of quantitative structure reactivity correlations. Naphthenes adsorb more strongly than aromatics on metals; the reverse is observed on acid sites (Marin et al., 1982; Van Trimpont et al., 1988). The value of the adsorption constant that appears in the rate law thus depends on the implied type of catalytic site. Preliminary experiments using pure component and simple mixture feedstocks showed that the metal-catalyzed hydrogenation and dehydrogenation reactions were fast enough to make the reactions on the acid site rate controlling. Thus, adsorption on the acid site was considered to be more important, and literature values for the adsorption constants were used for each class of compound (Van Trimpont et al., 1988).

7.3 MODEL DEVELOPMENT

The naphtha reforming modeling at the pathways level has seven reaction families. Various reaction rules summarized in Table 7.2 have been applied to constrain the reaction network based on literature and experience. The specifics and rules

TABLE 7.1
Reaction Matrices for Naphtha Reforming Reactions at the Pathways Level

Reaction Family		Reaction Matrix

Dehydrocyclization
Test: the string H—C—C—C—C—C—H is required

$$
\begin{array}{c}
C \\ C \\ H \\ H
\end{array}
\begin{bmatrix}
0 & -1 & 1 & 0 \\
-1 & 0 & 0 & 1 \\
1 & 0 & 0 & -1 \\
0 & 1 & -1 & 0
\end{bmatrix}
$$

Hydrocracking, hydrogenolysis, and dealkylation
Test: the string C—C is required

$$
\begin{array}{c}
C \\ C \\ H \\ H
\end{array}
\begin{bmatrix}
0 & 1 & -1 & 0 \\
1 & 0 & 0 & -1 \\
-1 & 0 & 0 & 1 \\
0 & -1 & 1 & 0
\end{bmatrix}
$$

Isomerization (paraffin and naphthene)
Test: the string C—C—C—H is required

$$
\begin{array}{c}
C \\ C \\ C \\ H
\end{array}
\begin{bmatrix}
0 & 0 & 1 & -1 \\
0 & 0 & -1 & 1 \\
1 & -1 & 0 & 0 \\
-1 & 1 & 0 & 0
\end{bmatrix}
$$

Aromatization
Test: a naphthenic ring is required

$$
\begin{array}{c}
C \\ H \\ C \\ H \\ C \\ H \\ C \\ H \\ C \\ H \\ C \\ H
\end{array}
\begin{bmatrix}
0 & -1 & 1 & 0 & 0 & 0 & 0 & 0 & 0 & 0 & 0 & 0 \\
-1 & 0 & 0 & 1 & 0 & 0 & 0 & 0 & 0 & 0 & 0 & 0 \\
1 & 0 & 0 & -1 & 0 & 0 & 0 & 0 & 0 & 0 & 0 & 0 \\
0 & 1 & -1 & 0 & 0 & 0 & 0 & 0 & 0 & 0 & 0 & 0 \\
0 & 0 & 0 & 0 & 0 & -1 & 1 & 0 & 0 & 0 & 0 & 0 \\
0 & 0 & 0 & 0 & -1 & 0 & 0 & 1 & 0 & 0 & 0 & 0 \\
0 & 0 & 0 & 0 & 1 & 0 & 0 & -1 & 0 & 0 & 0 & 0 \\
0 & 0 & 0 & 0 & 0 & 1 & -1 & 0 & 0 & 0 & 0 & 0 \\
0 & 0 & 0 & 0 & 0 & 0 & 0 & 0 & 0 & -1 & 1 & 0 \\
0 & 0 & 0 & 0 & 0 & 0 & 0 & 0 & -1 & 0 & 0 & 1 \\
0 & 0 & 0 & 0 & 0 & 0 & 0 & 0 & 0 & 0 & 0 & -1 \\
0 & 0 & 0 & 0 & 0 & 0 & 0 & 0 & 0 & 1 & -1 & 0
\end{bmatrix}
$$

of each important reforming reaction family in the catalytic reforming pathways are explained below.

7.3.1 Dehydrocyclization

Paraffins can convert to naphthenes via the dehydrocyclization (DHC) reaction. The DHC reaction involves the breakage of two C—H bonds and the formation of one C—C bond and one H—H bond. Only four atoms change connectivity during this

TABLE 7.2
Reaction Rules for Catalytic Reforming Model at Pathways Level

Reaction Family	Reaction Rule
Dehydrocyclization	Formation of five-membered ring naphthenes only

Cracking (hydrocracking and hydrogenolysis)	Cracking of multiple bonds not allowed
	Hydrocracking of n-paraffins not allowed
	Hydrogenolysis of isoparaffins not allowed
	Cracking of C# < C5 not allowed

| Isomerization (paraffin and naphthene) | Not allowed for multiple bonds or aromatic bonds |

| Aromatization | Aromatization of all six-membered ring naphthenes only |

| Dealkylation | Dealkylation of aromatic compounds only |

reaction. Table 7.1 summarizes the reaction matrix, which can be applied to all the paraffins and isoparaffins that meet the specific tests and rules described below.

A number of tests and rules were applied to each species before its graph was permuted and operated with the reaction matrix. The rules are summarized in Table 7.2. (Joshi et al., 1999) One test was applied to all paraffins and isoparaffins before they could be considered to undergo the DHC reaction. The molecule was required to have at least five carbon atoms connected to each other in a chain. The rule used in the current model was based on Callender's observation (Callender and Meerbott, 1976) that the DHC of paraffins or isoparaffins using a commercial reforming catalyst (i.e., Pt on chlorided alumina) resulted in mostly five-membered naphthene rings. Thus, for DHC, only five-membered rings were allowed to be formed from direct DHC.

The literature shows that the DHC rate increases with increases in the paraffin carbon number (Sivasanker and Padalkar, 1988). Callender and Meerbott, (1976) studied the bifunctional mechanism and reached the conclusion that both metal and acid sites were necessary for the DHC reaction. To the extent that the formation of a carbenium ion intermediate on the acid site is rate controlling, its heat of formation (i.e., ΔH_f of the carbenium ion) could be a reasonable candidate for QSRC/LFER. Often a simple structural property will suffice. For example, the greater the degree of branching of a carbenium ion is, the greater its stability is (Gates, 1992). Also, the heat of reaction in the formation of the members of a homologous series of carbenium ions decreases with increases in carbon number. Application of the LFER concepts to heterogeneous acid-catalyzed processes has been successfully demonstrated in the literature (Chapter 3).

The mechanistic insights about the acid site chemistry guided the division of this reaction family into four subfamilies. The cyclization rate depends on the type of the reactant carbenium ion: n-hexane cyclization to methylcyclopentane is faster than n-pentane to cyclopentane because of the formation of a secondary attacking carbenium ion in hexane. Also, the rate depends on the type of carbenium ion formed in the cyclization reaction (secondary, tertiary, etc.). Thus, the reaction family can be further divided into five subfamilies based on the type of reactant and product carbenium ions in the reaction: tertiary → secondary, tertiary → primary, secondary → secondary, secondary → primary, and primary → primary.

7.3.2 HYDROCRACKING

The hydrocracking reaction leads to yield loss but concentrates the reformate in high-octane species. The rate of this reaction is comparable to that of the DHC reaction. The extent of naphthene cracking is considerably less than paraffin cracking since naphthenes are rapidly converted to aromatics. Here we focus only on paraffins. The bifunctional metal-acid catalyzed hydrocracking reaction mechanism involves carbenium ion chemistry: the paraffin is dehydrogenated on the metal site to an olefin, which is in turn protonated on the acid site to give a carbenium ion. The carbenium ion then undergoes isomerization and β-scission to give cracked products.

The reaction matrix of the hydrocracking reaction involves breakage of two C—C bonds and formation of two C—H bonds as summarized in Table 7.1. The test for this reaction includes a search of a C—C bond where both carbons have at least one hydrogen substituent. This allows the formation of an olefin intermediate. The algorithm therefore looks for a C(H)—C(H) string. The rules used in the model are summarized in Table 7.2. Multiple bonds are not allowed to undergo this reaction due to their stabilities.

Branched compounds are primary products, while C_4-cracking products are secondary compounds (Marin et al., 1982; Van Trimpont et al., 1988; Sivasanker and Padalkar, 1988). Based on this result and due to the unfavorable stability of primary carbenium ions, it was determined that n-paraffins were not allowed to hydrocrack directly to lower carbon products; only isoparaffins can crack at the branch points that allow for the formation of a secondary and tertiary carbon upon β-scission.

Preliminary kinetics analysis showed that the rate constant for hydrocracking correlated with the heat of formation of a carbenium ion in the β-scission step on the acid site. The β-scission reaction in the current model thus depends both on the type of the reactant carbenium ion and on the type of carbenium ion formed in the reaction. A branched octane such as 2,2,4-trimethylpentane cracks much faster than n-octane because of higher number of tertiary carbons as well as the generation of a tertiary carbenium ion after cracking. Hence, this reaction family can be further divided into four subfamilies based on the type of the reactant and product carbenium ions: type A: tertiary → tertiary, type B1: secondary → tertiary, type B2: tertiary → secondary, and type C: secondary → secondary.

7.3.3 HYDROGENOLYSIS

Hydrogenolysis is the cracking on the metal sites. Paraffins, isoparaffins, naphthenes, and the side chains of alkyl aromatics are all susceptible to this reaction. The mechanism of this reaction involves adsorption of the saturated compound and breakage of the C–C bond on the metal site. Gault et al. (1981) summarized various mechanisms for the hydrogenolysis reaction on a Pt-alumina catalyst over the Pt 111 surface. They showed that the percentage of platinum could influence the selectivity of hydrogenolysis. The current model is based on conditions where nonselective hydrogenolysis is dominant.

The reaction matrix for the hydrogenolysis reaction is the same as for hydrocracking (Table 7.1). The test for this reaction includes a search for a C—C bond in the molecule. The rules used for this reaction are summarized in Table 7.2. As in hydrocracking, multiple bonds are not allowed to undergo this reaction. Also, paraffin and isoparaffin molecules with fewer than five carbon atoms are not allowed to undergo this reaction.

Gault et al. (1981) observed that the increase in the branching of the carbons decreases the rate of reaction. This reaction is thus limited to only nonbranched C—C bonds and is used to produce the light ends distribution (C1 and C2 especially) in our modeling.

7.3.4 PARAFFIN ISOMERIZATION

Paraffin isomerization, on a bifunctional catalyst, is caused by the reaction mechanism involving the dehydrogenation of the paraffin to an olefin on the metal site and subsequent protonation and isomerization on the acid site. The net result is summarized in the pathway levels reaction matrix shown in Table 7.1.

Isomerization involves the breakage of one C—C and one C—H bond and the formation of one C—H and one C—C bond. Only one to three isomerizations were considered in the modeling. Thus, the molecule was required to have at least four carbon atoms in series. The rules used for this reaction are summarized in Table 7.2.

Isomerization involving multiple bonds was prohibited because of the steric requirements.

Paraffin isomerization can be further divided into subfamilies based on the mechanistic understanding of the reaction on the acid site (the methyl shift and PCP (Protonated Cydo-Propane) isomerization mechanism). Parallel to the methyl shift in the mechanism, at the pathways level model, we have a subfamily called Isom-A, which only shifts the branch in a molecule. Parallel to the PCP isomerization steps, we have Isom-B, which changes the degree of branching. It is expected that Isom-A would be faster than Isom-B. Isom-B can be broken down further as needed into four subfamilies, namely straight to mono-branched, mono- to di-branched, di- to tri-branched, and tri- to tetra-branched.

7.3.5 NAPHTHENE ISOMERIZATION

Naphthene isomerization sometimes is called ring expansion and contraction. This reaction transforms a five-membered naphthene ring into a six-membered naphthene ring, and vice versa. Since six-membered rings are converted rapidly to aromatics, the ring expansion mechanism provides a pathway for the conversion of five-membered ring naphthenes to aromatics.

The mechanism of naphthene isomerization is similar to that of paraffin isomerization. A cyclopentane ring dehydrogenates to a cyclopentene ring, the subsequent protonation of which allows for the carbon shift. The reaction matrix is identical to that for paraffin isomerization, as in Table 7.1. The test for this reaction includes a search for a C—C—C—H string. The rules used for this reaction are summarized in Table 7.2.

7.3.6 DEHYDROGENATION (AROMATIZATION)

Naphthene dehydrogenation or aromatization is fast. Dehydrogenation predominantly occurs on the metal site, and, in practice, the equilibrium lies far toward aromatics. The dehydrogenation reaction matrix is shown in Table 7.1, and a model reaction is shown in Figure 7.1. The test for this reaction includes search for a six-membered naphthenic ring. No specific rules are needed for this reaction.

Under reforming conditions, the aromatization reaction is so fast that it is very difficult to measure its kinetics experimentally. Ritchie et al. (1966) studied the dehydrogenation of various substituted cyclohexanes and observed that the activation energy for the dehydrogenation reaction decreased with increases in branching on the cyclohexane reactant. This suggests that the activation energy can be correlated with the degree of substitution. In this model, the rate of aromatization was simply assumed to be fast enough to be in virtual equilibrium.

7.3.7 DEALKYLATION

Dealkylation of aromatics and naphthenes occurs in the reforming process, but to a considerably smaller extent than the other primary reactions. The reactivities of these reactions increase sharply with an increase in carbon number. The dealkylation reaction matrix is shown in Table 7.1, and a model reaction is shown in Figure 7.1. The test for this reaction includes a search for a ring. No specific rules are needed for this reaction.

7.3.8 COKING

The coking formation in naphtha reforming can be modeled to occur from the reactions of coke precursors, such as the naphthalene from feeds with a heavy tail, either present in the feed or created from the chemical reactions. From the modeling point of review, a suggested reaction route from the literature was used in this work: a concerted dealkylation and dehydrogenation of five substituted cyclopentanes, namely, cyclopentane (5N5), methylcyclopentane, ethylcyclopentane, propylcyclopentane, and butylcyclopentane, will produce cyclopentadiene. The stoichiometry of this reaction is R-N5 → RH + H2 + cyclopentadiene. Then, the Diels-Alder condensation reaction of cyclopentadiene will produce naphthalene, the coke precursor. Finally, the bimolecular condensation reaction of naphthalene and two moles of hydrogen will lead to a four-ring fused aromatic that donates coke.

7.4 AUTOMATED MODEL BUILDING

The reaction matrices, rules, and kinetic correlations described above were used to construct various reforming models. The model building algorithm (NetGen) used to generate the reforming models Figure 7.2 (Joshi et al., 1999) classified

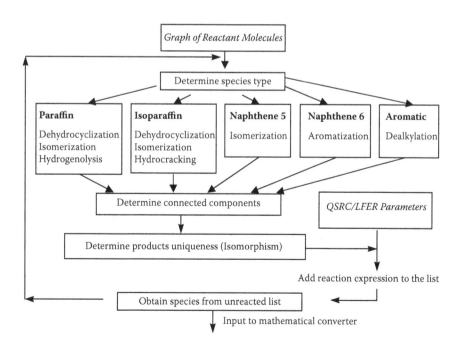

FIGURE 7.2 Algorithm for pathways level naphtha reforming model builder.

the molecules into the families of paraffins, isoparaffins, five-membered ring naphthenes, six- membered ring naphthenes, and aromatics. The reactions of each family member were separately handled by incorporating an additional filter on the type of reactions allowed for each compound class. The products obtained were checked for structural isomorphism and then were identified for the compound type. The current version of the reforming model builder contains the information about the reaction matrices and rules for all reforming reactions with the flexibility to change them. A library of naphtha molecules comprising the connectivity information and the thermodynamic properties was also built. The reactivity indices required for various reactions were calculated on the fly, and the rate expressions were written in a file as an input to the converter (EqnGen). The mathematical converter then generated the rate equations in the classical Langmuir–Hinshelwood–Hougen–Watson (LHHW) form. The reactor model was solved for the case of nonisothermal plug flow with molar expansion.

Many candidate models were built automatically by using the above tools by applying different sets of the reaction rules in various "what-if" scenarios. Since the analysis of experimental results at the reforming conditions suggested that olefins were present in low concentrations and were in a virtual equilibrium with paraffins, olefins were eliminated in the model except for cyclopentadiene, as it is used in the coking reaction.

For a C14 naphtha, the direct application of the algorithm will create too many isomers and an unmanageable situation. This is because the size of the models increases as 2^N, N being the carbon number. A model that was fully molecularly explicit up to C11 had 650 compounds with 2500 reactions. Thus, the following strategy was applied to extend the current naphtha-reforming model-building capability to C14 range.

The model builder was first guided to execute all reactions to the full details up to C8. From C9 to C11, there was a gradual diminution of the level of detail. The model builder was guided to create a decreasing number of molecular components to represent all isomers within a class, with the explicit accounting of the reaction path degeneracy (RPD). For C12 to C14, the model builder was guided to create only a few representative isomers to depict all of the isomers within a carbon number class.

The RPD is defined as the number of ways the same reaction can take place, which was specifically counted by the model builder and appended as a constant multiplier in the rate expression. For example, the number of ways for the cracking of butane to two ethane molecules is half the number for cracking of butane to methane and propane. The RPD also explains the reason bigger molecules (higher carbon numbers) crack faster than smaller ones because there are more bonds to break.

7.5 THE MODEL FOR C14 NAPHTHA REFORMING

The final C14 model contains 147 components and 587 reactions. The first 69 components are composed of all C8 molecules, and the others are representatives of C9 to C14 and three molecules involved in the coking model. Table 7.3

TABLE 7.3
Characteristics of the C14 Naphtha Reforming Model

Species	Number	Reactions	Number
Hydrogen	1	Dehydrocyclization	148
n-Paraffins	14	Paraffin isomerization	206
Isoparaffins	59	Hydrocracking	89
Naphthenes	56	Hydrogenolysis	20
Aromatics	14	Naphthene isomerization	74
Olefin (cyclopentadiene)	1	Aromatization	28
Coke precursor (naphthalene)	1	Dealkylation	15
Coke	1	Coking	7
Total number of species	147	Total number of reactions	587

summarizes the breakdown of molecular types and reaction types. The model was built in only 18 CPU seconds and can be solved in 2 CPU seconds on a Pentium Pro PC (200 MHz). These reactions are modeled as reversible reactions.

All the reaction families were constrained to have the same frequency factor, and the activation energy was correlated with the heat of reaction. The rate constant parameters for naphthene isomerization and aromatization were held fixed at high values to guarantee equilibrium between naphthenes and aromatics. The naphthene isomerization rate constant was held constant due to the lack of analytical information about six-membered naphthenes, which had very low concentrations in the product slate.

The adsorption parameters in the model were expressed by the Arrhenius formalism for each class of compound. Thus, a single expression was used for all paraffins, naphthenes, and aromatics. To evaluate equilibrium constants, values for the free energy of formation (ΔG_f) for each species were obtained from the literature (Stull, 1969; Alberty et al., 1994a,b,c).

An exponential decay function was used to represent the activity drop due to coking, as shown in Equation 7.2.

$$\varphi_R = \exp(-\alpha_R C_C) \qquad (7.2)$$

7.6 MODEL VALIDATION

The C14 model was solved for a three-bed plug flow reactor (PFR) with a defined external temperature profile. The initial guess values for the activation energies and frequency factors were obtained from the literature (Marin et al., 1982; Van Trimpont et al., 1988; Sivasanker et al., 1988; Dumesic et al., 1993). Due to the high correlation of the parameters (As and Es), the activation energies were constrained to be between 10 to 80 Kcal/mole, and the frequency factors between

0.0001 to 10 $(gmoles/m^3)^{1-n}(gm-catalyst.h)^{-1}$. These constraints were based on the need for physically meaningful values for the rate parameters.

The C14 model was then optimized by using the GREG optimization algorithm (Stewart et al., 1992) to tune the QSRC/LFER parameters. The molecular product composition was further organized into the following categories to match with experimental and pilot data: H2; C1; C2; C3; P4; n-, 1br-, mbr-P5, 5N, 6N, A for C6-C11; P, N, A for C12-14; coke precursor. The objective function was the sum of the squares of the difference between all the observed and predicted concentrations of the above lumps. Each molecule or lump was associated with a weighting parameter to ensure that the final objective function was unbiased. All the data sets were optimized together to give a sufficient degree of freedom to the optimization program. GREG was also used to obtain confidence limits on the parameters along with the information about the covariance matrix, which, in turn, gave information about the sensitivity and correlation of parameters.

Figure 7.3 (Joshi et al., 1999) shows the parity plots between the tuned model predictions of set of naphtha reforming pilot plant data for various feeds. As can be seen from Figure 7.3, the model is feedstock independent. Both the low-yield products and the high-yield products were well predicted. The absolute yield values

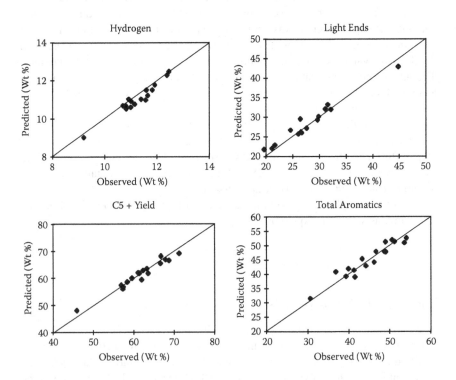

FIGURE 7.3 Parity plots of naphtha reforming model results for various feedstocks at various operation conditions.

for hydrogen and light ends matched well, in turn making the predictions of C5+ yield well. Figure 7.3 also shows a good prediction of aromatics in the product stream.

Various insights about reforming chemistry can be drawn from model results, especially by breaking down the main reaction families into subfamilies from the mechanistic insights. The type A (methyl shift) paraffin isomerization reaction was found to have lower activation energy than all the type B (branching) reactions. The activation energies for hydrocracking reactions of type A ($3° \rightarrow 3°$) and type B1 ($2° \rightarrow 3°$) are much lower than those of type B2 ($3° \rightarrow 2°$) and type C ($2° \rightarrow 2°$). A similar observation was made for the DHC reaction: the greater the branching is, the faster the reaction is.

7.7 SUMMARY AND CONCLUSIONS

The graph-theoretical approach and the reaction family concept were successfully applied to the molecule-based kinetic modeling of the bifunctional heterogeneous catalytic naphtha reforming process. The reaction matrices and reaction rules were reviewed for each reaction family and tabulated for easy reference. The automated model building capabilities were extended to the naphtha reforming at the pathways level and enable us to build molecular models fast and accurately for a variety of what-if scenarios.

The C14 naphtha catalytic reforming model with 147 species and 587 reactions was built and optimized and provided excellent parity between the predicted and experimental yields for a wide range of feedstocks and operating conditions. This demonstrated the feedstock-independent nature of the developed detailed kinetic model.

From the optimized model results, various chemical insights could be drawn by analyzing the subfamilies divided via mechanistic understanding of naphtha reforming. The inclusion of subfamilies in the DHC, isomerization, and hydrocracking reactions made it possible to incorporate mechanistic ideas in the pathways level model. The stability of the carbenium ions involved in the reaction greatly affects the rate of reaction, which we will further exploit in the mechanistic modeling of paraffin hydrocracking in the next chapter.

REFERENCES

Alberty, R.A. and Gehrig, C.A., Standard chemical thermodynamic properties of alkane isomer groups, *J. Phys. Chem. Ref. Data,* 13(4), 1173–1197, 1984a.

Alberty, R.A. and Ha, Y.S., Standard chemical thermodynamic properties of alkylcyclo-pentane isomer groups, alkylcyclohexane isomer groups, and combined isomer groups, *J. Phys. Chem. Ref. Data,* 14(4), 1107–1132, 1984b.

Alberty, R.A., Standard chemical thermodynamic properties of alkylbenzene isomer groups, *J. Phys. Chem. Ref. Data,* 14(1), 177–192, 1984c.

Alberty, R.A., Standard chemical thermodynamic properties of alkene isomer groups, *J. Phys. Chem. Ref. Data,* 14(3), 803–820, 1984d.

Antos, G.J., Aitani A.M., and Parera, J.M., Eds., *Catalytic Naphtha Reforming,* Marcel Dekker, New York, 1995.

Brandenberger, S.G., Callender, W.L., Meerboot, W.K., Mechanisms of methylcyclopentane ring-opening over platinum-alumina catalyots, *J Catal.,* 42(2), 282–287, 1976.

Dumesic, J.A., Rudd, D.F., Aparicio, L.M., Rekoske, J.E., and Trevino, A.A., *The Microkinetics of Heterogeneous Catalysis,* American Chemical Society, Washington, DC, 1993.

Gates, B.C., *Catalytic Chemistry,* John Wiley & Sons, New York, 1992.

Gault, F.G., Mechanisms of skeletal isomerization of hydrocarbons on metals, *Adv., Catal.,* 30, 1–95, 1981.

Joshi, P.V., Molecular and Mechanistic Modeling of Complex Process Chemistries, Ph.D. dissertation, University of Delaware, Newark, 1998.

Joshi, P.V., Klein, M. T., Huebner, A.L., Leycrle, R.W., Automated Kinetic modeling of catalytic reforming at the reaction pathways level, *Rev. Process Chem. Engin.,* 2, (3), 169–193 (1999).

Kmak, W.S., A Kinetic Simulation Model of the Powerforming Process, paper presented at AIChE Natl. Meet. Preprint, Houston, TX, 1971.

Marin, G.B. and Froment, G.F., Reforming of C6 hydrocarbons on a Pt-Al2O3 catalyst, *Chem. Eng. Sci.,* 37(5), 759–773, 1982.

Mudt, D.R., Hoffman, T.W., and Hendan, S.R., The Closed-Loop Optimization of a Semi-Regenerative Catalytic Reforming Process, AIChE paper-w51, Houston, 1995.

Ramage, M.P., Graziani, K.R., Schipper, P.H., Krambeck, F.J., and Choi, B.C., KINPTR (Mobil's kinetic reforming model): a review of Mobil's industrial process modeling philosophy, *Adv. Chem. Eng.,* 13, 193, 1987.

Ritchie, A.W. and Nixon, A.C., Dehydrogenation of Monocyclic Naphthenes Over a Platinum on Alumina Catalyst Without Added Hydrogen, Div. of Petro. Chem. Preprint, ACS, New York, Sept 11–16, 1966.

Sivasanker, S. and Padalkar, S.R., Mechanism of dehydrocyclization of n-alkanes over platinum-alumina catalyst, *Appl. Catal.,* 39, 123–126, 1988.

Stewart, W.E., Caracotsios, M., and Sorensen, J.P., Parameter estimation from multiresponse data, *AIChE Journal,* 38(5), 641–650, 1992.

Stull, C.R., *The Chemical Thermodynamics of Organic Compounds,* John Wiley & Sons, New York, 1969.

Van Trimpont, P.A., Marin G.B., and Froment, G.F., Reforming of C7 hydrocarbons on a sulfided commercial Pt/Al2O3 catalyst, *Ind. Eng. Chem. Res.,* 27, 51–57, 1988.

8 Mechanistic Kinetic Modeling of Heavy Paraffin Hydrocracking

8.1 INTRODUCTION

This chapter describes the application of the Kinetic Modeler's Toolbox (KMT) to the development of kinetic models for catalytic hydrocracking—a flexible heterogeneous bifunctional catalyzed process used extensively in the petroleum industry to convert heavy oils into lighter and more valuable products. Developing a model with detailed kinetic information requires modeling of the chemistry at either the pathways level or the mechanistic level. Compared with pathways level modeling, mechanistic modeling is accompanied by a tremendous increase in the number of species, the number of reactions, and the number of associated rate constants in the governing network because of the explicit accounting of all reaction intermediates such as the surface species. This can render mechanistic modeling very tedious. However, mechanistic models have more fundamental rate constant information and thus can be easily extrapolated not only to various operating conditions, but also to various catalysts in a family.

8.2 MECHANISTIC MODELING APPROACH

The main goal of mechanistic modeling is to obtain fundamental information that can be extrapolated over a wide range of operating conditions, feedstocks, and catalyst systems. This chapter extends the graph-theoretic modeling approach for heterogeneous mechanistic models. Various mechanistic models have been built for paraffins ranging from C16 to C80. All the models incorporate mechanistic chemistry on the bifunctional hydrocracking catalyst including both the metal function for hydrogenation and dehydrogenation and the acid function for isomerization and cracking. The model development process will be described.

Mechanistic models are much larger than pathways level models. Thus, the issues of rigorous representation of mechanistic chemistry (including stable molecules and unstable surface intermediates) for the large reaction network and reduction of the rate constants into physically significant parameters are more important and critical for mechanistic models.

To this end, it is still useful to realize that much of the complexity is statistical or combinatorial and that the large demand for reactions and rate parameters in the mechanistic model can be handled by lumping the reactions involving similar transition states into one reaction family, the kinetics of which are constrained to follow a quantitative structure reactivity relationship (QSRC) or linear free energy relationship (LFER). Due to the modeling of the chemistry at the mechanistic level, the organization of the reaction into families is more meaningful and fundamental.

The paraffin hydrocracking reaction mixtures are grouped into a few species types (paraffins, olefins, ions, and inhibitors such as NH_3), which, in turn, react through a limited number of reaction families on the metal (dehydrogenation and hydrogenation) and the acid sites (protonation, hydride-shift, methyl-shift, protonated-cyclopropane (PCP) isomerization, β-scission, and deprotonation). As a result, a small number of formal reaction operations can be used to generate hundreds of reactions.

The division of all the hydrocracking reactions into a small number of reaction families with an associated reaction matrix has been automated to exploit the repetitive nature of the operations. Similar to the pathways level modeling, graph-theoretic concepts have been exploited to automate the mechanistic hydrocracking reaction network generation.

Figure 8.1 shows the mechanistic reactions for paraffin hydrocracking via the dual site (metal site and acid site) mechanism. Thus, similar to the pathways level model, a reaction matrix for each reaction family was used to generate all possible mechanistic reactions. The reaction matrices for the mechanistic reaction families in paraffin hydrocracking are summarized in Table 8.1.

The rate constant information was organized with a QSRC/LFER for each reaction family shown in Equation 8.1,

$$E^* = E_o^* + \alpha \Delta H_{rxn} \qquad (8.1)$$

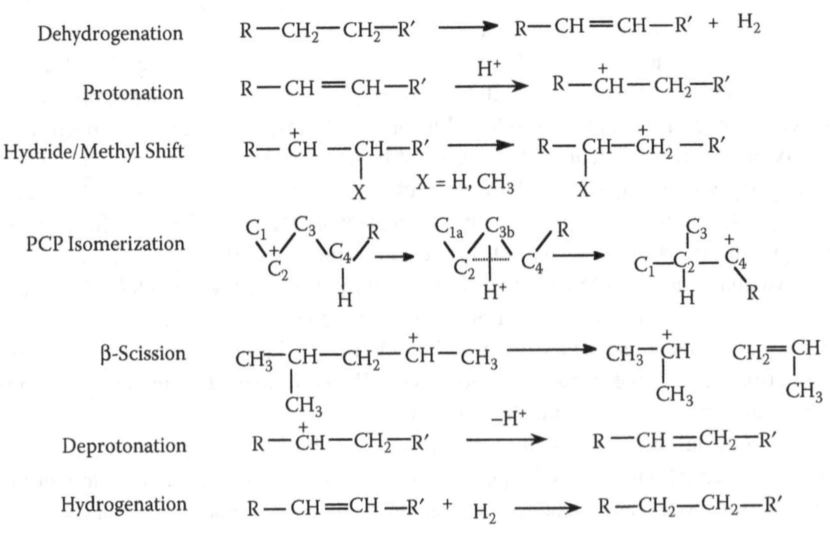

FIGURE 8.1 Paraffin hydrocracking reaction families at the mechanistic level.

TABLE 8.1

Reaction Matrices for Paraffin Hydrocracking Reactions at the Mechanistic Level

Reaction Family	Reaction Matrix
Dehydrogenation Test: the string C—C is required.	C: 0 1 −1 0 C: 1 0 0 −1 H: −1 0 0 1 H: 0 −1 1 0
Hydrogenation Test: the string C=C is required.	C: 0 1 −1 0 C: 1 0 0 −1 H: −1 0 0 1 H: 0 −1 1 0
Protonation Test: the string C=C is required.	C: 0 −1 1 C: −1 0 0 H+: 1 0 0
Deprotonation Test: the string C+—C is required.	C+: 0 0 1 C: 0 0 −1 H: 1 −1 0
Hydride shift and methyl shift Test: the string C+—C—X is required in the carbenium ion (X = H or C).	C+: 0 0 1 C: 0 0 −1 X: 1 −1 0
PCP isomerization Test: the string C+—C—C is required in the carbenium ion.	C+: 0 0 1 0 C: 0 0 −1 1 C: 1 −1 0 −1 X: 0 1 −1 0
β-Scission Test: the string C+—C—C is required in the carbenium ion.	C+: 0 1 0 C: 1 0 −1 C: 0 −1 0

the Polanyi relationship, which relates the activation energy of each reaction to its heat of reaction. A single frequency factor is assumed for each reaction family. The use of QSRC/LFER has been demonstrated by Korre et al. (1994) for heterogeneous mechanistic models, Russell (1996) for hydrocracking,

Watson et al. (1996, 1997) and Dumesic et al. (1993) for catalytic cracking, and Mochida and Yoneda (1967s) for dealkylation and isomerization.

The remainder of this chapter will describe the division of the paraffin hydrocracking reactions into mechanistic families with a unique reaction matrix operator for each reaction family. The reaction rules and QSRCs used will be discussed for each reaction family. The technical specifications and the iteration process used to find the optimum subset of the mechanistic model will also be discussed.

8.3 MODEL DEVELOPMENT

In this section, we will first discuss the reaction mechanism of paraffin hydrocracking and the specifications for each reaction family. Then we will discuss the automated model building algorithm and the kinetic QSRC/LFERs used to organize rate parameters. Finally, we will present the developed C16 paraffin hydrocracking model at the mechanistic level.

8.3.1 REACTION MECHANISM

The mechanism of hydrocracking reactions over bifunctional catalysts has been investigated extensively. The mechanism is essentially the carbenium ion chemistry of catalytic cracking coupled with that of the dehydrogenation and hydrogenation reactions. The initial reactions in hydrocracking are similar to those in catalytic cracking, but the presence of excess hydrogen and a hydrogenation component in the catalyst results in hydrogenated products and inhibits some of the secondary reactions such as secondary cracking and coke formation.

The mechanism of paraffin hydrocracking over bifunctional amorphous catalysts was studied in detail in the 1960s. Based on the pioneering work of Mills et al. (1953) and Weisz (1962), a carbenium ion mechanism was proposed, similar to the catalytic cracking plus additional hydrogenation and skeletal isomerization. More recent studies of paraffin hydrocracking over noble metal-loaded, zeolite-based catalysts have concluded that the reaction mechanism is similar to that proposed earlier for amorphous, bifunctional hydrocracking catalysts (Langlois and Sullivan, 1970; Weitkamp, 1975).

Figure 8.1 summarizes the reaction elementary steps used to describe the mechanism of paraffin hydrocracking over a bifunctional catalyst:

1. Dehydrogenation of paraffins to olefins on metal sites
2. Protonation of olefins to carbenium ions on acid sites
3. Carbenium ion hydride shift on acid sites
4. Carbenium ion methyl shift on acid sites
5. Protonated-cyclopropane (PCP) intermediate mediated branching of carbenium ion on acid sites
6. Carbenium ion cracking through β-scission on acid sites
7. Deprotonation of carbenium ions to olefins on acid sites
8. Hydrogenation of olefins to paraffins on metal sites

The model building algorithm incorporates the above reaction mechanism. It can classify the species into classes containing paraffins, olefins, and carbenium ions as well as H+ ions. It also identifies the branching degree of each species (1, 2, 3, and more) and the type of carbenium ions (primary, secondary, and tertiary). The specifics of each reaction family and rules are described below.

8.3.2 REACTION FAMILIES

8.3.2.1 Dehydrogenation and Hydrogenation

In the typical hydrocracking process, the reactions occurring on the metal sites of the catalyst include mainly the dehydrogenation and hydrogenation reactions. Very little hydrogenolysis occurs on the metal sites. The metal component of the catalyst dehydrogenates the paraffin reactants to produce reactive olefin intermediates, hydrogenates the olefins from cracking, and prevents catalyst deactivation by hydrogenating coke precursors.

Paraffin is adsorbed on the metal site and dehydrogenated to olefin, which then desorbs from the metal site and diffuses to the acid site. In the presence of the acid function, olefins can react via carbenium ion chemistry.

The mechanism of the dehydrogenation reaction involves the stripping of two hydrogen atoms by the metal component of the catalyst. The test for the dehydrogenation reaction involves a search of a C—C string in the molecule. Figure 8.2 shows the number of olefins as a function of the carbon number. It can be seen that the number of possible olefins increases almost exponentially with the carbon number, and even one paraffin can form thousands of olefins, hence inclusion of all possible olefins and their reactions would generate an enormously large model. In order to develop a mechanistic model that is of reasonable size to facilitate rapid solution on available computational resources, certain rules were used for this reaction. They are summarized in Table 8.2. All n-paraffins were allowed to undergo dehydrogenation reactions at all sites, whereas all isoparaffins were allowed to undergo dehydrogenations only at the C—C bonds β to the branch. This rule was based on the relative rate of reactions of these olefins on the acid site.

The dehydrogenation and hydrogenation of paraffins on hydrocracking catalysts, especially the commonly used noble metal-loaded zeolite catalysts, is very fast, and thermodynamic equilibrium concentrations are rapidly reached. Thus, the rate parameters of this reaction were constrained by the thermodynamic equilibrium constants.

8.3.2.2 Protonation and Deprotonation

Protonation transforms an olefin into a carbenium ion. This reaction is much faster than other acid site reactions and is close to equilibrium under commercial operating conditions. The protonation reaction involves attack of H+ at a C=C bond. Only three atoms change connectivities during the reaction. The reaction matrix (connectivity changer) is shown in Table 8.1.

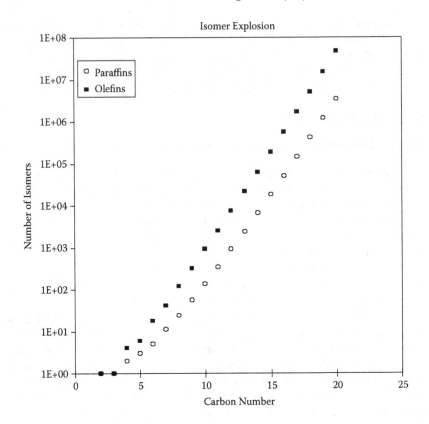

FIGURE 8.2 Number of paraffins and olefins as a function of carbon number.

Mechanistic understanding from the literature was used to synthesize the tests and rules. The deprotonation reaction, which converts carbenium ions to olefins, involves breakage of a C—H bond to give H+ and an olefin. The deprotonation test requires a connected C+ and C atom. The rules are summarized in Table 8.2. Primary carbenium ions are not allowed to form from protonation due to their thermal–chemical instabilities.

8.3.2.3 Hydride and Methyl Shift

Hydride and methyl shifts are responsible for changes in the position of carbenium ions. The net effect is generally to create a stable ion (e.g., a tertiary ion) from a less stable ion (e.g., a secondary ion). The methyl shift can also change the location of a branch position, which creates isomers.

The rate of hydride shift is considered to be much faster than alkyl shift due to the ease of moving H+ compared to the alkyl group. The hydride shift reaction test requires a C+—C—H string in the molecule; that for the methyl shift reaction

TABLE 8.2
Reaction Rules for Mechanistic Paraffin Hydrocracking Reaction Families

Reaction Family	Reaction Rules
Dehydrogenation and hydrogenation	Dehydrogenation was allowed everywhere on n-paraffins and only β to the branch of isoparaffins. Formation of di-olefins not allowed.
Protonation and deprotonation	Primary carbenium ions were not allowed to form.
Hydride shift and methyl shift	Primary carbenium ions were not allowed to form. Migration to a stable ion or branched ions. Number of reactions allowed is a function of branch number.
PCP isomerization	Primary and methyl carbenium ions were not allowed to form. PCP-isomerization increases either the number of branches or the length of the side chains. PCP isomerization to form vicinal branches was not allowed. Only methyl and ethyl branches were allowed; a maximum of three branches was allowed. Number of reactions allowed is a function of branch and carbon number.
β-Scission	Methyl and primary carbenium ions were not allowed to form.

requires a C+—C—(CH$_3$) string. The rules are summarized in Table 8.2. The reactions are allowed for all ions. The number of reactions allowed is constrained as a function of the number of branches on the ions. This provides the proper spectrum of isomers and keeps the number of species and reactions manageable.

8.3.2.4 PCP Isomerization

The isomerization reaction is postulated to proceed via a protonated cyclopropane (PCP) intermediate with the charge delocalized over the ring (Brouwer et al., 1972). The rate of this reaction is slower than that of the hydride and methyl shift. This reaction is further categorized into two types, isomA and isomB, depending on the identity of the bond to be broken in the three-membered ring intermediate.

The rules for this reaction have a dramatic effect on the size of the generated model. The test for this reaction requires a C+—C—C string in the molecule. The final set of rules used for the model building are summarized in Table 8.2, and the reaction is shown in Figure 8.1. As is the case for the hydride shift and methyl shift, the isomerization reaction is allowed for all paraffins and isoparaffins, and the number of reactions is constrained as a function of the number of carbons and branches on the ions to provide the proper spectrum of isomers and to keep the number of species and reactions manageable.

8.3.2.5 β-Scission

The β-scission reaction is one of the key carbon number-reducing reactions for isoparaffins. The rate of this reaction is dependent on the acidity of the catalyst. β-scission can lead to the formation of tertiary and secondary carbenium ions, but no primary ions are formed. Several β-scission mechanisms have been suggested for the cracking of branched secondary and tertiary carbenium ions, as summarized in Table 8.3 (Martens et al., 1986).

Type A β-scission that converts a tertiary carbenium ion to another tertiary carbenium ion is the most likely to occur. The reaction rates decrease in the order of A, B1, B2, and C. Each type of reaction requires a minimum number of carbon atoms in the molecule and a certain kind of branching in order to occur. The proposed β-scission mechanisms suggest that paraffins may undergo several isomerizations until a configuration is attained that is favorable to β-scission.

The test for this reaction requires a C+—C—C string. The rules are summarized in Table 8.2. Unstable species, such as methyl and primary carbenium ions, are not allowed to form from this reaction.

8.3.2.6 Inhibition Reaction

The inhibition effects of nitrogen compounds in the hydrocracking process can be accounted for by the dynamic reduction of acid sites by introducing the protonation and deprotonation of the Lewis base on the acid sites. For example,

---

TABLE 8.3
β-Scission Mechanisms for Carbenium Ion Conversion over Bifunctional Hydrocracking Catalyst

Type	Min C#	Ions Involved	Rearrangement
A	8	Tert → Tert	
B1	7	Sec → Tert	
B2	7	Tert → Sec	
C	6	Sec → Sec	

the following ammonia inhibition reaction can be included in the reaction network:

$$NH_3 + H^+ \leftrightarrows NH_4+ \tag{8.2}$$

With this inhibition reaction included in the reaction network, the above inhibition protonation and deprotonation will compete with hydrocarbon protonation and deprotonation and thus reduce the available number of acid sites for hydrocarbons, as we can see from Equation 8.3 since the total ion concentration is conserved.

$$H_0^+ = H^+ + \sum_{i=1}^{N} R^+ + NH_4^+ \tag{8.3}$$

8.3.3 AUTOMATED MODEL BUILDING

The reaction matrices in Table 8.1, the reaction rules in Table 8.2, the QSRC/LFER correlations of Equation 8.1, and the automated model building algorithm of Figure 8.3 were used to construct various paraffin hydrocracking mechanistic kinetic models from C8 to C24 on the computer. Figure 8.4 depicts a representative reaction network of paraffin hydrocracking at the mechanistic

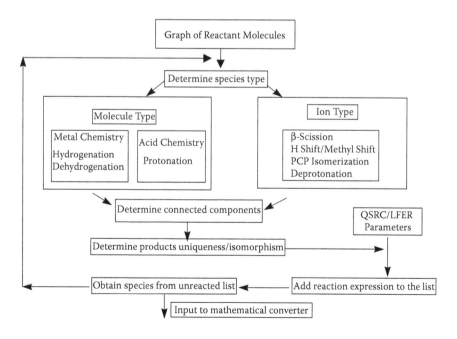

FIGURE 8.3 Algorithm for automated paraffin hydrocracking at the mechanistic modeling level.

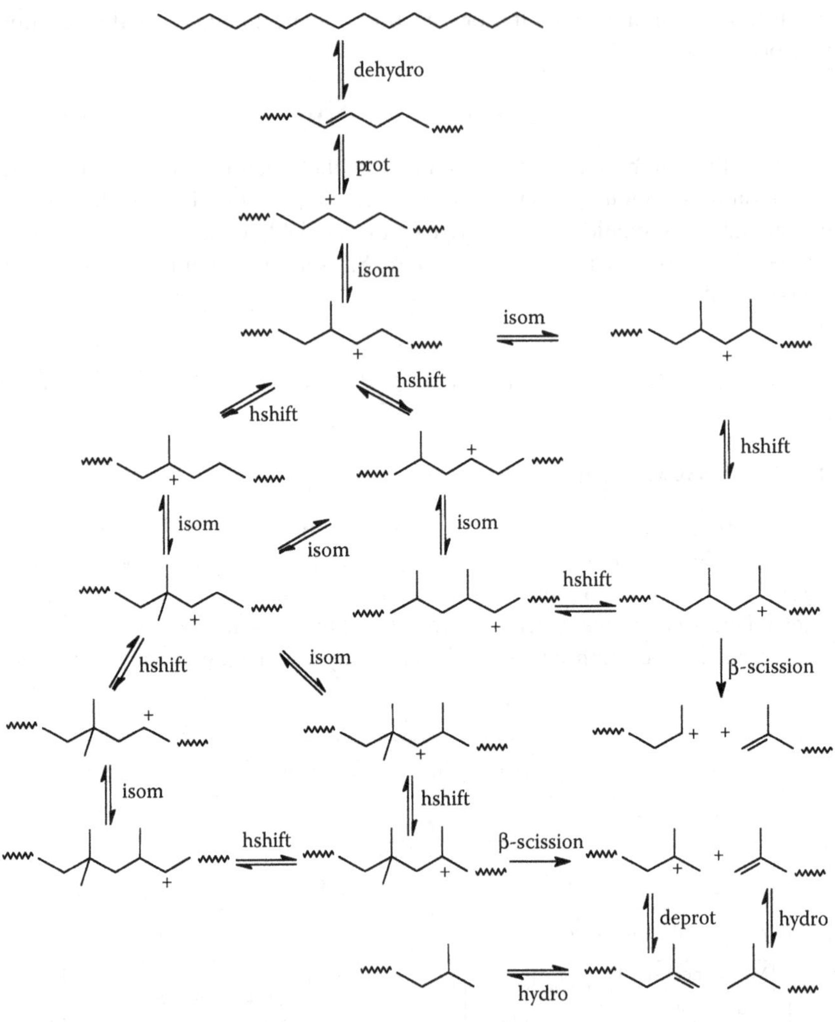

FIGURE 8.4 Representative reaction network of paraffin hydrocracking at the mechanistic level.

level — how a large paraffin molecule goes through various reactions including dehydrogenation, protonation, H/Me-shift, various isomerizations, and forms the configuration to carry out the β-scission and breaks down to smaller ions and olefin and then finally deprotonates and hydrogenates to smaller paraffin molecules. It shows that the reaction pathways leading to the formation and subsequent cracking of a tertiary carbenium ion are preferred in the complete reaction network. These insights and lessons learned from building smaller models then guided the extension of the model building process to the heavier paraffin hydrocracking model up to as high as C80, as we will discuss in detail later.

The model building algorithm used to generate the mechanistic hydrocracking models, as shown in Figure 8.3, similar to that of the pathways level models, classified the molecular graphs of the reactants into the families of molecules and ions. The molecules were further filtered to undergo specific metal and acid chemistries. For example, paraffins were only allowed to undergo dehydrogenation reactions on the metal sites, whereas olefins were allowed to hydrogenate on the metal site and protonate on the acid sites.

The current version of the mechanistic paraffin hydrocracking model builder has a database of all reaction matrices and rules for all mechanistic paraffin hydrocracking reactions, both metal and acid chemistries, with the flexibility to change them. A detailed library of paraffins, olefins, and ion intermediates consisting of their connectivity and thermodynamic information was built using the MOPAC computational chemistry package. The reactor-balance equations were generated for a plug flow reactor (PFR) with molar expansion.

8.3.4 KINETICS: QUANTITATIVE STRUCTURE REACTIVITY CORRELATIONS

The rate constants within each reaction family were described in terms of a reaction family-specific Arrhenius A factor and the Polanyi relationship parameters that related the activation energy to the enthalpy change of reaction, as shown in Equation 8.1.

The Polanyi relationship and the Arrhenius expression can be combined to represent the rate constant k_{ij}, where i denotes the reaction family and j denotes the specific reaction in the family:

$$k_{ij} = A_i \exp(-(E_{o,i} + \alpha_i * \Delta H_{rxn,j})/RT) \tag{8.4}$$

The acidity of the catalyst was captured by a single parameter $\Delta H_{stabilization}$ (Dumesic et al., 1993), signifying the relative stabilization of the H^+ ion compared to other carbenium ions. Since the reactions on the acid site are rate controlling, this was a useful way to separate the catalyst property (acidity) in the rate constant formalism and is shown in Equation 8.5.

$$k_{ij} = A_i \exp(-(E_{o,i} + \alpha_i * (\Delta H_{rxn,j} - \Delta H_{stabilization}))/RT) \tag{8.5}$$

Each reaction family could be described with a maximum of three parameters (A, Eo, α). Procurement of a rate constant from these parameters required only an estimate of the enthalpy change of reaction for each elementary step. In principle, this enthalpy change of reaction amounted to the simple calculation of the difference between the heats of formation of the products and reactants. However, since many model species, particularly the ionic intermediates and olefins, are without experimental values, the MOPAC computational chemistry package (Stewart, 1989) was used to estimate the heat of formations "on the fly." The organization of the rate constants into QSRCs reduced the number of model parameters from $O(10^3)$ to $O(10)$.

8.3.5 THE C16 PARAFFIN HYDROCRACKING MODEL AT THE MECHANISTIC LEVEL

A C16 hydrocracking model with 465 species and 1503 reactions was built automatically in only 14 CPU seconds on a Pentium Pro 200 PC. The corresponding PFR model with molar expansion was then generated automatically and solved in 76 CPU seconds once through. Table 8.4 summarizes the characteristics of the C16 model.

The species distribution in Table 8.4 shows there were more intermediates, namely olefins and ions, than the final product molecules, namely paraffins. The reaction distribution in Table 8.4 shows that each molecule reactant needs to go through various rearrangements to form the right configuration before β-scission. In this model, all type A, B1, and B2 β-scission reactions were allowed; the type C reaction was ignored in the final model after optimization with experiment results, where we found the C-type cracking was insignificant. Hydride shift, methyl shift, and isomerization reactions were restricted for hydrocarbons having more than eight carbon atoms. All n-paraffins and selected isoparaffins (each considered as a lump) were used to represent the portion of the feedstock larger than C9. This not only helped keep the model size reasonable, but also resulted in the inclusion of components with different reactivities in the feedstock, where detailed characterization was not available.

TABLE 8.4
Characteristics of the C16 Paraffin Hydrocracking Model

Species	Number	Reaction	Number
Molecule		Dehydrogenation	233
Hydrogen	1	Hydrogenation	233
Pariffins	64	Deprotonation	328
Olefin	233	Protonation	328
Ion		Hydride and methyl shift	168
H-ion	1	PCP isomerization	174
Carbenium ions	165	β-Scission	37
Inhibitor	1	Inhibition	2
Total number of species	465	Total number of reactions	1503

8.4 MODEL RESULTS AND VALIDATION

The C16 models were optimized with pilot plant data using the GREG optimization program (Stewart et al., 1992). This nonlinear regression program solved the equations for the different species yields for both the stable molecular species (paraffins) and the unstable intermediate species (both olefins and ions). The objective function was the square of the difference between predicted and experimental yields weighted by the experimental standard deviation as shown in Equation 8.6.

$$F = \sum_{i=1}^{M} \sum_{j=1}^{N} \left(\frac{y_{ij}^{\text{model}} - y_{ij}^{\text{exp}}}{\bar{\omega}_j} \right)^2 \tag{8.6}$$

where i is the experiment number, j is the species or lump number, and ωj is the experimental measurement deviation.

The A factors were obtained from literature (Watson et al., 1996, 1997) and were held constant during optimization. The A factors for reactions on the metal site were constrained between $1 \leq A_j \leq 20$, and the α's in the QSRC formalism were held at 0.5 based on guidelines from the literature (Watson et al., 1996, 1997; Dumesic et al., 1993). The E_o's for reversible reactions (hydrogenation/dehydrogenation, protonation/deprotonation, isomerization/re-isomerization) were constrained by the relation between activation energies and heat of reaction shown in Equation 8.7.

$$E_{backward} - E_{forward} = \Delta H_{rxn} \tag{8.7a}$$

$$E_{oj,forward} = E_{oj,backward} \tag{8.7b}$$

$$\alpha_{j,forward} = 1 - \alpha_{j,backward} \tag{8.7c}$$

Thus, only the E_o's and one catalyst stabilization parameter were optimized by matching with the experimental data.

The model was optimized with the lumped data from experiments (the unconverted C16, the monobranched and multibranched C16, and all the cracked products). Then the model was used to predict the carbon number distribution and iso-to-normal ratio along the carbon numbers. Figure 8.5 shows the parity plots for the C16 paraffin hydrocracking for the lumped observations on the left and the carbon number distribution on the right at various operation conditions (T, P, LHSV, and NH_3). The agreement between the predicted and experimental data sets for all the high- and low-yield compounds is good, and all predictions

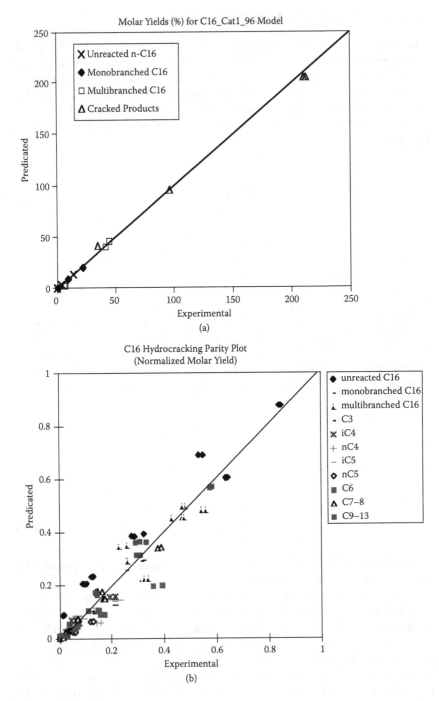

FIGURE 8.5 Parity plots for C16 paraffin hydrocracking model at various operation conditions (T, P, LHSV, and inhibition).

are within the experimental error. The good prediction validates the fundamental nature of the model.

Various insights were obtained from the optimization results for this set of paraffin hydrocracking experiments on a commercial bifunctional Pt-Zeolite catalyst. The results showed that the reaction rates were ranked as follows: A-type cracking > H/methyl shift > PCP isomerization > B1-type cracking > B2-type cracking >> C-type cracking. Thus, the C-type cracking can be neglected in our case; all the cracking products come from A-type cracking of tri-branched ions or B-type cracking of di-branched ions. This explains why the cracking by β-scission is much more likely to occur with tri-branched and di-branched isomers than with monobranched ones (Martens et al., 1986). Wherever several reaction pathways are possible, the one leading to the formation and subsequent cracking of a tertiary carbenium ion is preferred. Furthermore, the cracking of smaller paraffins via β-scission is less likely to occur, which explains their high yields even at high conversions. From the product molecular structure point of view, PCP isomerization always leads to branching, A-type cracking always leads to branched isomers, and B-type cracking always leads to normal or branched isomers.

In the experiments, practically no methane and ethane formation was observed. This confirmed our modeling assumption of allowing no primary ions because of their instability compared to more stable secondary and tertiary ions. This eliminated the formation of methane and ethane via the carbenium ion mechanism in the reaction network. This also partially explains why the long-chain paraffins tend to crack at or near the center.

The reaction mechanism in which secondary carbenium ions are isomerized to more stable tertiary ions prior to cracking, as well as the high rate of H-shift to the tertiary carbenium ion, explains the high iso-to-normal ratio for paraffins in the product. The iso-to-normal ratio in the product paraffins increases with decreasing reaction temperature because at higher temperatures, the cracking rate of isoparaffins increases faster than that of the n-paraffins.

The ammonia inhibition reduces not only the cracking activity but also the iso-to-normal ratio in product paraffins because of its partial neutralization of the acid sites on the hydrocracking catalyst.

8.5 EXTENSION TO C80 MODEL

Applying the same model building algorithm and reaction rules to C16 to C80 would easily use up computer memory long before a reaction network could be completely built. This is because of the enormous number of isomers associated with high-carbon-number hydrocarbons. Moreover, the current analytical capability to identify these high-carbon-number isomers is limited. The higher the carbon number range, the more process engineers and chemists are interested in lumped rather than in individual molecular level detail. A molecular model with the detail of carbon number and branch number distribution level for C80 would serve modeling purposes quite well.

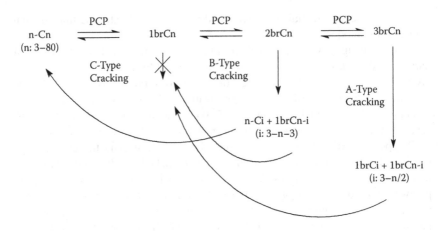

FIGURE 8.6 Paraffin hydrocracking reaction mechanism deduced reaction pathways at the carbon number and branch number level.

Various model building and control strategies have thus been exploited to construct a high-carbon-number model, including the stochastic rule-based algorithm. All the details can be found in Chapter 3.

By restricting the isomorphic criteria to the carbon and branch number level in the generalized isomorphism algorithm, the molecules with the same carbon number and branch number are lumped together. This allows construction of a molecule-based paraffin hydrocracking model at the pathways level for the complex feedstock up to C80.

Figure 8.6 summarizes the essential paraffin hydrocracking chemistry at the carbon and branch number level, which is reduced to several PCP isomerizations (normal and one, two, and three branch) and cracking (type A and type B) reactions. This captures all the key observations from the paraffin hydrocracking reaction mechanism: (a) paraffins go through isomerizations before cracking; (b) PCP isomerization always leads to branching; (c) A-type cracking always leads to branched isomers; (d) B-type cracking always leads to normal or branched isomers; and (e) all the cracking products come from A-type cracking of tri-branched isomers or B-type cracking of di-branched isomers. The C-type cracking can be neglected in our case.

A C80 paraffin hydrocracking model with the above reaction pathways with 306 species and 4671 reactions was automatically generated on the computer. This model can be further tuned to capture the carbon number distribution and iso-to-normal ratio in the paraffin hydrocracking process.

8.6 SUMMARY AND CONCLUSIONS

Graph-theoretic concepts and the QSRC formalism were used to construct a paraffin hydrocracking mechanistic model builder incorporating bifunctional catalysis with metal and acid functions. The complete automated approach was successfully demonstrated for a C16 paraffin hydrocracking model with 465 species and 1503 reactions at the mechanistic level.

The reaction family concept was exploited in the molecule-based kinetic modeling of paraffin hydrocracking by representing the reactions with various reaction families incorporating the metal function (dehydrogenation/hydrogenation) and the acid function (protonation/deprotonation, H/Me-shift, PCP isomerizations, and β-scission). The optimized C16 model provided excellent parity between the predicted and experimental yields for a wide range of operating conditions. This reinforces the fundamental nature (feedstock and catalyst acidity independent) of the rate parameters in the model.

Various insights were obtained from the optimization results of the detailed C16 model: (a) the skeletal isomerizations precede the cracking reactions; (b) PCP isomerization always leads to branching, A-type cracking always leads to branched isomers, and B-type cracking always leads to normal or branched isomers; and (c) all the cracking products normally come from A-type cracking of tri-branched isomers or B-type cracking of di-branched isomers.

The generalized isomorphism algorithm has been applied at the carbon and branch number level to reduce the complexity and explosion of modeling heavy paraffin hydrocracking. A C80 paraffin hydrocracking model with 306 species and 4671 reactions at the pathways level reduced from the fundamental chemistry and reaction mechanism has thus been developed.

REFERENCES

Broadbelt, L.J., Stark, S.M., and Klein, M.T., Computer generated pyrolysis modeling: on-the-fly generation of species, reactions and rates, *Ind. Eng. Chem. Res.,* 33, 790–799, 1994a.

Broadbelt, L.J., Stark, S.M., and Klein, M.T., Computer generated reaction networks: on-the-fly calculation of species properties using computational quantum chemistry, *Chem. Eng. Sci.,* 49, 4991–5101, 1994b.

Broadbelt, L.J., Stark, S.M., and Klein, M.T., Computer generated reaction modeling: decomposition and encoding algorithms for determining species uniqueness, *Comput. Chem. Eng.,* 20, 2, 113–129, 1996.

Bhore, N.A., Klein, M.T., and Bischoff, K.B., The Delplot technique: a new method for reaction pathway analysis, *Ind. Eng. Chem. Res.,* 29, 313–316, 1990.

Brouwer, D.M. and Hogeveen, H., Electrophilic Dubstitutions at aukenes and in alkylcarbonium ions, *Prog. Phys. Org Chem.,* 9, 179, 1972.

Coonradt, H.L. and Garwood, W.E., Mechanism of hydrocracking reactions of paraffins and olefins, *I. & E. C. Proc. Des. Dev.,* 3, 38, 1964.

Dumesic, J.A., Rudd, D.F., Aparicio, L.M., Rekoske, J.E., and Trevino, A.A., *The Microkinetics of Heterogeneous Catalysis,* American Chemical Society, Washington, DC, 1993.

Gates, B.C., *Catalytic Chemistry,* John Wiley & Sons, New York, 1992.

Hou G., Mizan, T.I., and Klein M.T., Computer-Assisted Kinetic Modeling of Hydroprocessing, Symposium on Catalysis in Fuel Processing and Environmental Protection, ACS Preprint, 1997.

Korre, S.C., Quantitative Structure/Reactivity Correlations as a Reaction Engineering Tool: Applications to Hydrocracking of Polynuclear Aromatics, Ph.D. thesis, University of Delaware, Newark, 1994.

Langlois, G.E. and Sullivan R.F., Chemisty of hyrocracking, *Adv. Chem. Ser.*, 97, 38, 1970.

Martens, J.A., Jacobs, P.A., and Weitcamp, J., attempts to rationalize the distribution of hydrocracked products: 1. Quantitative description of the primary hydrocracking modes of long chain paraffins in open zeolites, *Appl. Catal.*, 20, 239, 1986.

Mills, G.A., Heinemann, H., Milliken, T.H., and Oblad, A.G., Naphtha reforming involves dual functional catalysts mechanism for reforming with these catalysts, *Ind. Eng. Chem.*, 45, 134, 1953.

Mochida, I. and Yoneda, Y., Linear free energy relationships in heterogeneous catalysis. I. Dealkylation of alkylbenzenes on cracking catalysts, *J. Catal.*, 7, 386–392, 1967a.

Mochida, I. and Yoneda, Y., Linear free energy relationships in heterogeneous catalysis. II. Dealkylation and isomerization reactions on various solid acid catalysts, *J. Catal.*, 7, 393–396, 1967b.

Mochida, I. and Yoneda, Y., Linear free energy relationships in heterogeneous catalysis. III. Temperature effects in dealkylation of alkylbenzenes on the cracking catalysts, *J. Catal.*, 7, 223–230, 1967c.

Neurock, M. and Klein, M.T., When you can't measure—model, *CHEMTECH*, 23(9), 26–32, 1993.

Read, C.J., Reactant and Catalyst Structure/Function Relationships in the Hydrocracking of Alkyl-Substituted Biphenyl Compounds, Ph.D. dissertation, University of Delaware, Newark, 1995.

Russell, C.L., Molecular Modeling of the Catalytic Hydrocracking of Complex Mixtures: Reactions of Alkyl Aromatic and Alkyl Polynuclear Aromatic Hydrocarbons, Ph.D. dissertation, University of Delaware, 1996.

Stewart, J. J.P., Semi-emprical molecular orbital methods; In: *Review of Computational chemisty*, Lipkowitz, K.B., Boyd, D.D., eds., New York, VCH Publishers, 1990.

Stewart, W.E., Caracotsios, M., and Sorensen, J.P., Parameter estimation from multiresponse data, *AIChE J.*, 38(5), 641–650, 1992.

Watson, B.A., Klein, M.T., and Harding, R.H., Mechanistic modeling of n-heptane cracking on HZSM-5, *Ind. Eng. Chem. Res.*, 35(5), 1506–1516, 1996.

Watson, B.A., Klein, M.T., and Harding, R.H., Catalytic cracking of alkylcyclohexanes: modeling the reaction pathways and mechanisms, *Int. J. Chem. Kinet.*, 29(7), 545, 1997.

Weisz, P.B., Polyfunctional heterogeneous catalysis, *Adv. Catal.*, 13, 137, 1962.

Weitkamp, J., *Hydrocracking and Hydrotreating*, ACS Symp. Series 20, Ward, J.W. and Qader, S.A., Eds., Washington, DC, 1975, p. 1.

9 Molecule-Based Kinetic Modeling of Naphtha Hydrotreating

9.1 INTRODUCTION

Hydrotreating is one of the most important and mature technologies in today's commercial refinery. The hydrotreating process removes objectionable materials from petroleum distillates by selectively reacting these materials with hydrogen in a catalyst bed such as CoMo or NiMo on Al_2O_3 at elevated temperatures. The objectionable materials include sulfur, nitrogen, olefins, and aromatics.

Hydrodesulfurization (HDS) process chemistry has been studied extensively over the past decades (Girgis and Gates, 1991; Topsøe et al., 1996; Whitehurst et al., 1998). With more and more stringent environmental regulations, interest in HDS continues. Low sulfur specifications have caused refiners to look at hydrotreating options, and thus more rigorous models are desired to improve the process. In this chapter, a rigorous molecular level modeling of naphtha hydrotreating is developed, with the focus on the HDS modeling. This is part of a wide-scale effort aimed at automated molecule-based kinetic modeling of gas oil hydroprocessing (Chapter 10).

HDS modeling so far has been focused on lumps of sulfur components, which were reacted via simple first-order or second-order kinetics. Deep HDS, however, requires a much more accurate modeling at the individual component level. The new paradigm tracks each molecule in the feed and product through the process. Developing such a model with detailed kinetic information requires modeling of the HDS chemistry at least at the reaction pathways level for each molecule in the process. Pathways models offer the promise of containing detailed kinetic information by the explicit inclusion of observable molecules and, hence, have the predictive capabilities of feedstock independence that lumped models lack. Various mechanistic insights into the chemistry can be incorporated into the reaction pathways too.

The following section describes the automated kinetic modeling approach with the associated reaction matrices and reaction rules. The next section discusses the hydrotreating process chemistry and reaction networks at the reaction pathways level with the focus on HDS for various sulfur-compound types including mercaptans, sulfides, disulfides, thiophenes (Ts), benzothiophenes (BTs), dibenzothiophenes (DBTs), and their alkyl and hydrogenated derivatives. Then the dual-site mechanism (σ site for direct desulfurization and τ site for hydrogenation and saturation on the catalyst surface) is incorporated, and the corresponding dual-site Langmuir–Hinshelwood–Hougen–Watson (LHHW) formalism is con-structed to describe the complex kinetics. Both the steric and electronic effects of the alkyl side chain of thiophenic compounds (T, BT, and DBT) are taken into account. The model building process is automated, and a rigorous molecule-based kinetic model for naphtha hydrotreating is thus developed. The optimized model shows very good agreement with experiments and provides useful quantitative insights.

9.2 MODELING APPROACH

As discussed in Chapters 2 and 3, much modeling complexity is statistical or combinatorial, and the large demand for reactions and rate parameters in the molecule-based kinetic models can be handled by organizing the reactions involv-ing similar transition states into one reaction family, the kinetics of which are constrained to follow the linear free energy relationship (LFER) or quantitative structure-reactivity correlations (QSRC).

Applying these ideas to naphtha hydrotreating starts with an examination of the feedstock. The reaction mixture can be grouped into a few compound classes (sulfur compounds, paraffins, olefins, naphthenes, aromatics, and nitrogen com-pounds), which react through a limited number of reaction families (desulfurization, sulfur saturation, olefin hydrogenation, aromatic saturation, and denitrogenation). As a result, a small number of reaction matrices can be used to generate the reaction network with hundreds of reactions, as discussed in Chapter 2. The reactions pathways in the hydrotreating process and the corresponding reaction matrices for all reaction families are summarized in Figure 9.1 and Table 9.1, respectively.

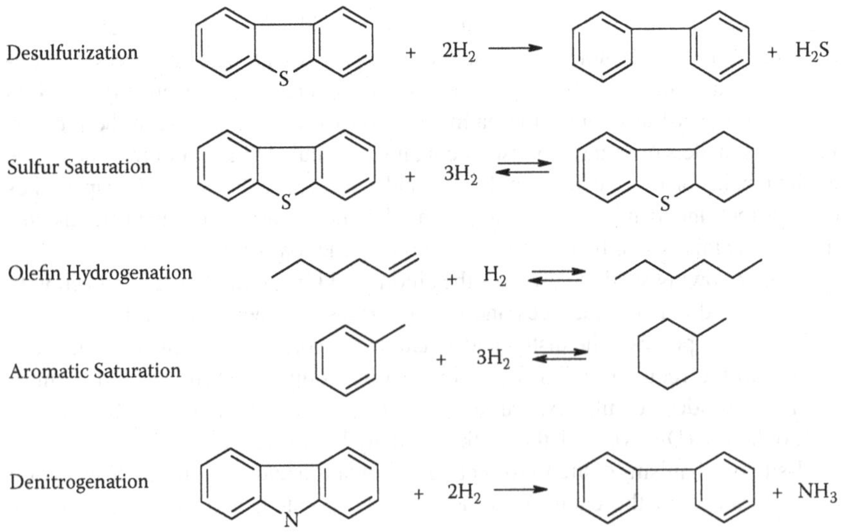

FIGURE 9.1 Hydrotreating reaction families at the pathways level.

The rate constant information was organized with a QSRC/LFER for each reaction family shown in Equation 9.1:

$$E^* = E^*_o + \alpha\Delta H_{rxn} \tag{9.1}$$

which relates the activation energy of each reaction to its heat of reaction. A single frequency factor was assumed for each reaction family. The use of QSRC/LFER for heterogeneous kinetic models has been demonstrated by

---

TABLE 9.1
Reaction Matrices for Hydrotreating Reactions at the Pathways Level

Reaction Family	Reaction Matrix							
Desulfurization (C—S case) Test: the string C—S is required	C	0	−1	1	0			
	S	−1	0	0	1			
	H	1	0	0	−1			
	H	0	1	−1	0			
Desulfurization (S—S case) Test: the string S—S is required	S	0	−1	1	0			
	S	−1	0	0	1			
	H	1	0	0	−1			
	H	0	1	−1	0			
Desulfurization (C—S—C case) Test: the string C—S—C is required	C	0	−1	0	1	0	0	0
	S	−1	0	−1	0	0	1	1
	C	0	−1	0	0	1	0	0
	H	1	0	0	0	−1	0	0
	H	0	0	1	−1	0	0	0
	H	0	1	0	0	0	0	−1
	H	0	1	0	0	0	−1	0
Sulfur saturation (2H case) or olefin hydrogenation Test: the string C=C is required	C	0	−1	1	0			
	C	−1	0	0	1			
	H	1	0	0	−1			
	H	0	1	−1	0			
Sulfur saturation (4H case) Test: The string C=C—C=C is required	C	0	−1	0	0	1	0	0	0
	C	−1	0	0	0	0	1	0	0
	C	0	0	0	−1	0	0	1	0
	C	0	0	−1	0	0	0	0	1
	H	1	0	0	0	0	−1	0	0
	H	0	1	0	0	−1	0	0	0
	H	0	0	1	0	0	0	0	−1
	H	0	0	0	1	0	0	−1	0

(Continued)

TABLE 9.1 (Continued)
Reaction Matrices for Hydrotreating Reactions at the Pathways Level

Reaction Family	Reaction Matrix

Sulfur saturation (6H case) or aromatic saturation
Test: an aromatic ring is required

C	0	-1	0	0	0	0	1	0	0	0	0	0
C	-1	0	0	0	0	0	0	1	0	0	0	0
C	0	0	0	-1	0	0	0	0	1	0	0	0
C	0	0	-1	0	0	0	0	0	0	1	0	0
C	0	0	0	0	0	-1	0	0	0	0	1	0
C	0	0	0	0	-1	0	0	0	0	0	0	1
H	1	0	0	0	0	0	0	-1	0	0	0	0
H	0	1	0	0	0	0	-1	0	0	0	0	0
H	0	0	1	0	0	0	0	0	0	-1	1	0
H	0	0	0	0	0	0	0	0	-1	0	0	0
H	0	0	0	0	1	0	0	0	0	0	0	-1
H	0	0	0	0	0	1	0	0	0	0	-1	0

Denitrogenation (C—N—C case)
Test: the string C—N—C is required

C	0	-1	0	1	0	0	0
N	-1	0	-1	0	0	1	1
C	0	-1	0	0	1	0	0
H	1	0	0	0	-1	0	0
H	0	0	1	-1	0	0	0
H	0	1	0	0	0	0	-1
H	0	1	0	0	0	-1	0

Korre (1995) and Russell (1996) for hydrocracking, Watson et al. (1996, 1997) and Dumesic et al. (1993) for catalytic cracking, and Mochida and Yoneda (1967a,b, and c) for dealkylation and isomerization. In addition, the steric hindrance effect of the alkyl side chains near the S atom were addressed as described later in this chapter.

9.3 MODEL DEVELOPMENT

In this section, we will first review the reaction families in the hydrotreating process. Then we will discuss the reaction kinetics used to organize all the reaction families and various sulfur compounds on the dual sites of the catalyst. Finally, we will discuss how to automate the model building process for naphtha hydrotreating.

9.3.1 REACTION FAMILIES

Hydrotreating modeling at the pathways level has five reaction families: desulfurization, sulfur saturation, olefin hydrogenation, aromatic saturation, and denitrogenation.

TABLE 9.2

Reaction Rules for Hydrotreating at Pathways Level

Reaction Family	Reaction Rule
Desulfurization	S is completely removed directly from the ring.
Sulfur saturation	Aromatic ring saturation proceeds in a ring-by-ring manner.
Olefin hydrogenation	Double bond not on the ring required to be hydrogenated.
Aromatic saturation	Aromatic ring saturation proceeds in a ring-by-ring manner.
Denitrogenation	N is completely removed directly from the ring.

The hydrotreating network building requires the reaction rules summarized in Table 9.2. The specifics of each reaction family in the hydrotreating pathways are explained below.

9.3.1.1 Reactions of Sulfur Compounds: Desulfurization and Saturation

Removing sulfur from the oil fraction is the main purpose of the hydrotreating process. Modeling this chemistry requires, first, classification of various sulfur compounds and then identification of their reactivities, reaction pathways, and mechanisms.

9.3.1.1.1 Classification of Sulfur Compounds

The petroleum feedstock mainly contains the following sulfur compound types: mercaptans (or thiols), sulfides, disulfides, Ts, BTs, DBTs, and their alkyl and hydrogenated derivatives. By combining several sequenced liquid chromatographic separations with gas chromatography–mass spectroscopy (GC–MS) and by using gas chromatography atomic emission detection (GCAED) for sulfur compounds, it has been possible to identify the majority of individual sulfur species in some fuels (Whitehurst et al., 1998). HDS reactivity depends critically on the molecular size and structure of the sulfur compounds. The mercaptans, sulfides, and disulfides generally have fast kinetics compared with the thiophenic compounds and are increasingly difficult to desulfurize from T to BT to DBT. The structures and carbon numbering of T, BT, and DBT are shown in Figure 9.2. The substituent groups α or adjacent to the S atom on thiophenic compounds generally retard HDS. While methyl groups distant from the S atom generally increase HDS activity — an effect attributed to increased electron density on the S atom — those adjacent to the S atom decrease reactivity due to the steric hindrance effect (Topsøe et al., 1996).

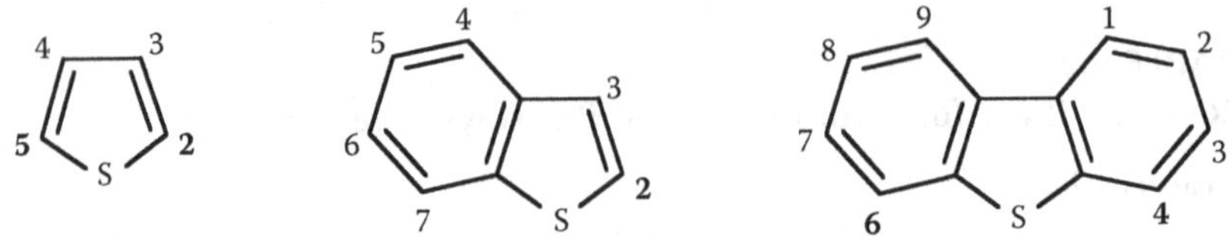

FIGURE 9.2 Thiophenes (T), benzothiophenes (BT), and dibenzothiophenes (DBT).

Table 9.3 identifies the "significant" positions on T, BT, and DBT, which are also shown in bold in Figure 9.2. The significant position means that if there is an alkyl substituent in that position, it will significantly affect the HDS reactivity of that compound due to the steric and electronic effects. For example, it has been found that 4,6-dimethyldibenzothiophene (4,6-DMDBT) is the most difficult to desulfurize and remains intact until the final stages of HDS processing of a light oil. The substituent groups on the significant and nonsignificant positions have different steric and electronic effects on HDS reactivities. As shown in Table 9.3, from the molecular modeling point of view, for example, we need at least two mono-methyl, three di-methyl, and three tri-methyl substituted molecular structures to account for this position difference for alkyl-DBTs up to C3 in the naphtha range, since we have classified the positions on the ring into two categories. Table 9.4 to Table 9.6 list the representative molecular structures of alkyl-Ts, alkyl-BTs, and alkyl-DBTs, as well as their rate adjustment factors compared to the parent molecule, T, BT, and DBT, respectively. From the molecule-based kinetic modeling perspective, these are the basic sets of thiophenic molecular structures we need to take into account to model the HDS reactions. It is also noteworthy that when the sulfur content is lowered from 0.20% to less than 0.05%, the chemistry of the HDS of naphtha or gas oils is essentially the chemistry of alkyl-DBTs. Though naphtha and gas oils initially contain mostly alkyl-Ts or alkyl-BTs, these are completely removed by the time 0.20% S is achieved (Whitehurst et al., 1998).

TABLE 9.3
Significant Positions of Thiophenic Compounds

Sulfur	Significant Position	Nonsignificant	Factor (electronic + steric)[a]	Representative Structure
T	2, 5	3, 4	$f_{1,T} \ll f_{2,t}$	2C1, 3C2, 2C3, 1C4
BT	2	3, 7, 4, 5, 6	$f_{1,BT} \ll f_{2,BT}$	2C1, 2C2, 2C3, 2C4
DBT	4, 6	1, 2,3, 7, 8, 9	$f_{1,DBT} \ll f_{2,DBT}$	2C1, 3C2, 3C3

[a]Alkyl position adjustment factor for rate constants. Subscript 1 denotes there is an alkyl chain at a significant position; 2 for nonsignificant position.

TABLE 9.4
Representative Alkyl-Ts and Their Rate Adjustments
with Respect to T

Alkyl-T	Representative Molecules	Rate Adjustment Factor
C1-T		f_1
		f_2
C2-T		f_1^2
		$f_1 f_2$
		f_2^2
C3-T		$f_1^2 f_2$
		$f_1 f_2^2$
C4-T		$f_1^2 f_2^2$

9.3.1.1.2   Reaction Pathways and Networks

The reaction pathways shown in Figure 9.3 were used to describe HDS chemistry (Girgis et al., 1991; Topsøe et al., 1996; Whitehurst et al., 1998). The mercaptans, sulfides, and disulfides can be easily desulfurized. T, BT, DBT, and their alkyl derivatives can go through either the sulfur saturation reaction (hydrogenation),

TABLE 9.5
Representative Alkyl-BTs and Their Rate Adjustments with Respect to BT

Alkyl-BT	Representative Molecules	Rate Adjustment Factor
C1-BT		f_1
		f_2
C2-BT		$f_1 f_2$
		f_2^2
C3-BT		$f_1 f_2^2$
		f_2^3
C4-BT		$f_1 f_2^3$
		f_2^4

TABLE 9.6
Representative Alkyl-DBTs and Their Rate Adjustments with Respect to DBT

Alkyl-DBT	Representative Molecules	Rate Adjustment Factor
C1-DBT		f_1
		f_2
C2-DBT		f_1^2
		$f_1 f_2$
		f_2^2
C3-DBT		$f_1^2 f_2$
		$f_1 f_2^2$
		f_2^3

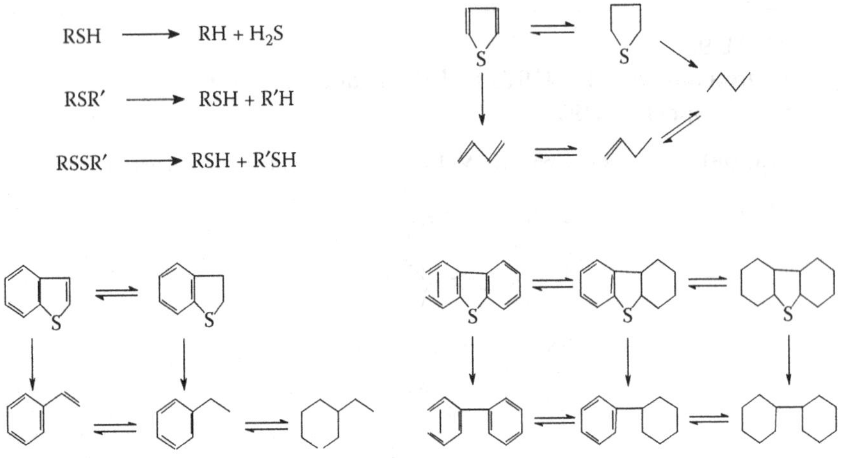

FIGURE 9.3 Sulfur compounds reaction pathways and network.

depicted as a reversible reaction, or the direct desulfurization reaction, which is depicted as an irreversible reaction in Figure 9.3.

For highly substituted DBTs, sulfur saturation prior to direct desulfurization is the major route to hydrocarbon production since aliphatic substituents on aromatic ring carbons adjacent to the sulfur atom impose severe steric hindrance of bonding to the catalyst surface and of the production of appropriate intermediate species, relative to the parent molecules. This scheme applies to all polyaromatic T-based HDS conversion processes; only the rate constants are different. With the more reactive DBTs, the rate of extraction of sulfur from the first saturated intermediate is so high that products derived from the fully saturated DBTs may not be observed. Similarly, the rate of saturation of single aromatic rings is generally much slower than rates of the other reactions; the dicyclohexanes are generally produced in trace amounts. However, as the alkyl-DBTs become less reactive, the importance of the above reactions increases and cannot be neglected.

The direct desulfurization reaction and the sulfur saturation reaction can be further classified into three subfamilies dependent upon the chemical bonds involved in the reaction. Direct desulfurization is further classified into the C—S case (as in mercaptans and sulfides), the S—S case (as in disulfides), and the C—S—C case (as in T, BT, DBT, and their alkyl and hydrogenated derivatives). The sulfur saturation is further classified into the 4H case (as in T and alkyl-Ts), the 2H case (as in BT and alkyl-BTs), and the 6H case (as in DBT and alkyl-DBTs). The reaction matrix for each case and its corresponding test are summarized in Table 9.1. The reaction rule for each reaction family is summarized in Table 9.2. For desulfurization, the sulfur is removed directly from the ring as H_2S, and simultaneous cleavage of two carbon–sulfur bonds from the thiophenic compound is assumed. For sulfur saturation, one ring at a time is saturated completely.

9.3.1.2 Olefin Hydrogenation

Although HDS is the most common of the hydrotreating reactions, olefin hydrogenation also proceeds quite rapidly. In this process, hydrogen is added to a double bond on an olefin or unsaturated naphthene; the corresponding hydrogenated compound is the product. This reaction is very fast compared with other hydrotreating reactions.

The reaction matrix for the olefin hydrogenation reaction involves breakage of a C–C bond and a H–H bond and formation of two C–H bonds as summarized in Table 9.1. The test for this reaction includes a search for a C=C bond. There are no special rules except that every olefin molecule is required to be hydrogenated in a hydrotreating process as summarized in Table 9.2.

9.3.1.3 Aromatic Saturation

Aromatic saturation is similar to olefin hydrogenation in that hydrogen is added to saturate the whole aromatic ring. For the single aromatic ring compounds in the naphtha range, such as the benzene and alkyl benzenes, the ring will be saturated as a whole with $3H_2$ or 6H.

The reaction matrix and test are summarized in Table 9.1, and the saturation 6H case and the rule are summarized in Table 9.2. However, for polynuclear aromatic compounds, two other types of aromatic saturation could happen: the 4H case for terminal aromatic rings, as in naphthalene, and the 2H case for middle aromatic rings, as in phenanthrene. The details will be discussed Chapter in 10, where we encounter these cases.

9.3.1.4 Denitrogenation

The nitrogen compounds that would normally be found in the feed of a hydrotreater can be classified into three categories: basic nitrogen compounds, which are generally associated with a six-membered ring, such as pyridine and quinoline; non-basic nitrogen compounds, which are generally associated with a five-membered ring, as in indole or carbazole; and the others such as anilines and amines, which are not common in crude oils but may be present in the hydrodenitrogenation (HDN) reaction network of the above nitrogen heterocyclics. The complexity of the nitrogen compounds makes denitrogenation even more difficult than desulfurization.

The HDN chemistry, from the modeling point of view, can be treated similarly to the HDS chemistry in the sense that two generic reaction families, nitrogen-compound saturation and direct denitrogenation, can be used to describe the chemistry. The details of HDN chemistry will be discussed in Chapter 10. Here, a simple lumped denitrogenation reaction will be used in the reaction network to represent the HDN chemistry.

The reaction matrix and test are summarized in Table 9.1, and the reaction rule is summarized in Table 9.2. Just like sulfur, nitrogen is removed from the ring directly; both carbon–nitrogen bonds cleave simultaneously, and N is removed from the ring as NH_3.

9.3.2 REACTION KINETICS

One view is that conventional HDS catalysts possess two very different types of catalyst sites, which contribute to the different reaction pathways described earlier. One induces the direct extraction of sulfur (desulfurization), and the other catalyzes ring hydrogenation (saturation).

Many researchers have observed that HDS reactions follow the LHHW kinetics (Girgis et al., 1991; Topsøe et al., 1996; Whitehurst et al., 1998). Equation 9.1 is the LHHW rate law used to model HDS kinetics. The classical dual-site mechanism (σ site for direct desulfurization and τ site for saturation on the catalyst surface in HDS processes) is utilized to implement the corresponding dual-site LHHW formalism. This rate law is derived from model compound studies by assuming the rate-determining surface reaction step between adsorbed reactants and two competitively adsorbed hydrogen atoms for both types of reaction.

$$r = \frac{fkK_{A,}K_{H,}[A][H_2]}{\left(1+\sum_i K_{i,}[I]+\sqrt{K_{H,}[H_2]}\right)^n} + \frac{fkK_{A,}K_{H,}([A][H_2]-[B]/K)}{\left(1+\sum_i K_{i,}[I]+\sqrt{K_{H,}[H_2]}\right)^n} \qquad (9.1)$$

In Equation 9.1, r is the reaction rate of a thiophenic compound, with the first term relating to the irreversible direct desulfurization and the second to the reversible ring saturation, $[I]$ is the concentration of the component, k is the rate constant, K_i is the adsorption constant of the component, K is the equilibrium constant, and n is the exponent of the inhibition term, in which 3 is assumed for HDS due to the reaction rate-determining step assumption. The rate adjustment factor, f, is introduced to account for the total steric and electronic effects of substituents on thiophenic compounds compared with their nonsubstituted parent molecule, as discussed in Section 9.3.1.1.2. The adsorption constants were obtained from the literature, and the following guidelines were followed: K_i (σ site): DBT > H_2S >> biphenyl >>> H_2; K_i (τ site): DBT > biphenyl >>> H_2. The inhibition or adsorption of H_2S is much greater on the catalyst site responsible for direct desulfurization rather than the hydrogenation site (Vanrysselberghe and Froment, 1996).

The above LHHW rate law takes into account the competition for the active sites on the catalyst surface between various species in the reaction stream including aromatics, olefins, and various types of sulfur compounds. Moreover, the inhibition to the HDS of the less reactive sulfur species from the H_2S and other hydrocarbons produced in the early stages of HDS are taken into account in the LHHW rate law.

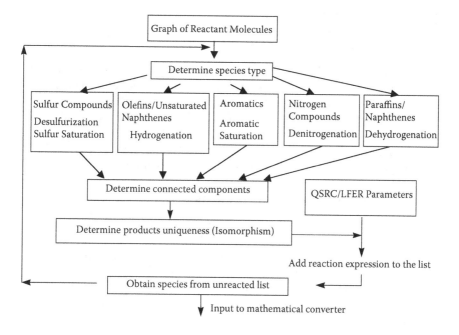

FIGURE 9.4 Algorithm for pathways level hydrotreating model builder.

9.3.3 AUTOMATED MODEL BUILDING

The reaction matrices, rules, and kinetic correlations described above were used to construct various hydrotreating models. The model building algorithm used to generate the hydrotreating models, as shown in Figure 9.4, classified the molecules into the families of sulfur compounds, paraffins (normal and isoparaffins), olefins (normal and iso-olefins), naphthenes (saturated and unsaturated), aromatics, and nitrogen compounds. The sulfur compounds are further classified as mercaptans, sulfides, disulfides, Ts, BTs, and DBTs. The reactions of each family member were separately handled by incorporating an additional filter on the type of reactions allowed for each compound class. The products obtained were checked for structural isomorphism and then were identified for the compound type. The current version of the hydrotreating model builder contains information about the reaction matrices and rules for all hydrotreating reactions with the flexibility to change them. A library of naphtha molecules comprising the connectivity information and the thermodynamic properties was also built. The reactivity indices required for various reactions were calculated on-the-fly, and the rate expressions were written in a file as an input to the converter. The mathematical equation converter then generated the rate equations in the LHHW form as discussed in the last section. The fixed-bed reactor model can then be solved easily.

9.4 RESULTS AND DISCUSSION

9.4.1 THE NAPHTHA HYDROTREATING MODEL

A detailed molecular level kinetic model for naphtha hydrotreating with 243 species and 437 reactions has been developed. The complete reaction model was built automatically in only 2 CPU seconds and solved in once-through mode in about 1 CPU second on an Intel Pentium II 333-MHz machine. The statistics of the model are summarized in Table 9.7. The model has 75 sulfur compounds (including H_2S) and 174 reactions involving sulfur compounds with 74 direct desulfurization reactions and 100 sulfur saturation reactions.

9.4.2 MODEL OPTIMIZATION AND VALIDATION

The model was solved for a fixed bed reactor and then optimized by using the GREG optimization algorithm (Stewart et al., 1992) to tune the QSRC/LFER parameters. The molecular product composition was further organized into the following categories to match experimental and pilot data: sulfur compounds, paraffins, isoparaffins, olefins, iso-olefins, naphthenes, unsaturated naphthenes, aromatics, and nitrogen compounds. The objective function was the sum of the squares of the difference between all the observed and predicted concentrations of the above lumps. Each molecule or lump was associated with a weighting parameter to ensure that the final objective function was unbiased. All the data sets were optimized together to give a sufficient degree of freedom to the optimization program. GREG was also used to obtain confidence limits on the parameters along with the information about the covariance matrix, which, in turn, gave information about the sensitivity and correlation of parameters.

TABLE 9.7
Statistics of the Naphtha Hydrotreating Model

Species	Number	Reactions	Number
Hydrogen	1	Olefin hydrogenation	74
Paraffins	9	Olefin dehydrogenation	74
Isoparaffins	14	Aromatic saturation	57
Naphthenes	42	Aromatic unsaturation	57
Unsaturated naphthenes	12	Sulfur saturation (2H)	9
Aromatics	50	Sulfur unsaturation (2H)	9
Olefins	14	Sulfur saturation (4H)	9
Iso-olefins	24	Sulfur unsaturation (4H)	9
H_2S	1	Sulfur saturation (6H)	32
Sulfur compounds	74	Sulfur unsaturation (6H)	32
NH_3	1	Desulfurization	74
Nitrogen compounds	1	Denitrogenation	1
Total number of species	243	Total number of reactions	437

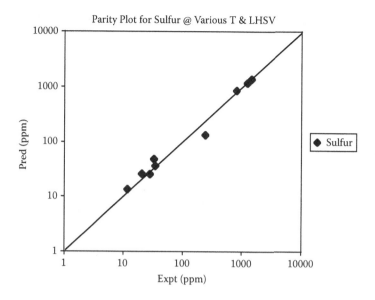

FIGURE 9.5a Parity plot of the naphtha hydrotreating model (for sulfur).

Figure 9.5 shows the parity plots between the tuned model predictions of a set of naphtha hydrotreating pilot plant data for various feeds at various operation conditions. As can be seen from Figure 9.5, the model matches the experimental data very well and can do an excellent job at a wide range of sulfur levels ranging from 10 ppm to 2000 ppm. Both the low-yield products, such as unsaturated naphthenes, n-olefins, iso-olefins, and n-paraffins, and the high-yield products, such as aromatics, naphthenes, and isoparaffins, were modeled very well. The aromatics did not convert much. This makes sense since most of the aromatics in the naphtha range are one-ring alkylbenzenes for which the ring saturation reaction is very slow. It is also noteworthy that the nitrogen compound lump was predicted extremely well in this model, even with just one lump and one reaction.

9.5　SUMMARY AND CONCLUSIONS

The graph-theoretical approach and the reaction family concept were successfully applied to the molecule-based kinetic modeling of the heterogeneous catalytic hydrotreating process. The reaction matrices and reaction rules were reviewed for each reaction family and tabulated for further reference. The automated model building capabilities were exploited and extended to the hydrotreating at the pathways level that enables us to build rigorous molecular models fast and accurately.

To model the HDS chemistry rigorously, it is necessary to incorporate at least all the representative molecular structures with substituents at both significant

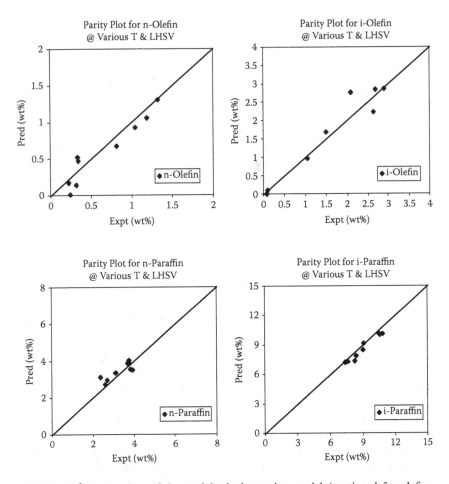

FIGURE 9.5b Parity plots of the naphtha hydrotreating model (continued for olefins, iso-olefins, paraffins, and isoparaffins).

and nonsignificant positions. It is also important to incorporate dual-site mechanisms and implement the corresponding LHHW formalism to take into account the inhibitions of various compounds in the process stream (especially the H_2S inhibition to the direct desulfurization sites). A structural approximation concept was introduced to account for the steric and electronic effects of substituents on thiophenic compounds to the kinetic reaction rates compared with their nonsubstituted parent molecule.

The developed naphtha hydrotreating model with 243 species and 437 reactions was built automatically and optimized and provided excellent parity between the predicted results and pilot plant data for a wide range of operating conditions.

Although we have developed this hydrotreating model for the naphtha range due to the availability of the experimental data, the modeling strategy and algorithms

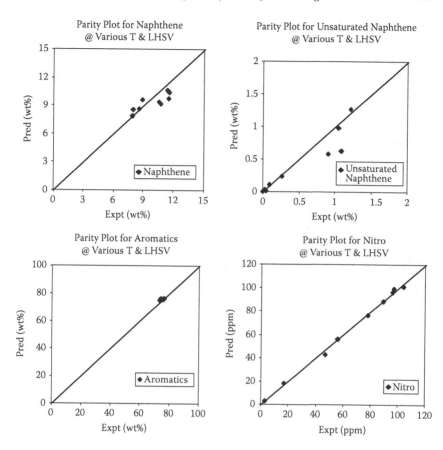

FIGURE 9.5c Parity plots of the naphtha hydrotreating model (continued for naphthenes, unsaturated naphthenes, aromatics and nitrogen compounds).

developed can be easily extended to the other oil fractions, such as gas oil, as we will discuss in Chapter 10.

REFERENCES

Dumesic, J. A., Rudd, D.F., Aparicio, L. M., Rekoske, J. E., and Trevino, A. A., The microkinetics of heterogeneous catalysis, *Amer. Chem. Soc.,* Washington DC., 1993.

Girgis, M.J. and Gates, B.C., Reactivities, reaction networks, and kinetic in high pressure catalytic hydroprocessing, *Ind. Eng. Chem. Res.,* 30, 2021, 1991.

Hou, G., Klein, M.T., Molecular Modeling of Gas Oil Hydrodesulfurization, paper presented at 2nd International Conference on Refinery Processes Proceedings, AIChE, Houston, TX, 1999.

Korre, S., Quantitative Structure/Reactivity Correlations as Reaction Engineering Tool: Applications to Hydrocracking of Polynuclear Aromatics, Ph.D. thesis, University of Delaware, Newark, 1995.

Ma, X., Sakanishi, K., and Mochida, I., Hydrodesulfurization reactivities of various sulfur compounds in diesel fuel, *Ind. Eng. Chem. Res.*, 33, 218–222, 1994.

Mochida, I. And Yoneda, Y., Linear free energy relationships in heterogeneous catalysis I. Dealkylation of alkylbenzenes on cracking catalysts., *J. Catal.*, 7, 386–392, 1967a.

Mochida, I. And Yoneda, Y., Linear free energy relationships in heterogeneous catalysis. II. Dealkylation and isomerization reactions on various solid acid catalysts, *J. Catal.*, 7, 393–396, 1967b.

Mochida, I. And Yoneda, Y., Linear free energy relationships in heterogeneous catalysis. III. Temperature effects in dealkylation of alkylbenzenes on the cracking catalysts, *J. Catal.*, 8, 223–230, 1967c.

Russell, C.L., Hydrocracking Reaction Pathways, Kinetics, and Mechanisms of n-Alkylbenzenes, M.S. thesis, University of Delaware, Newark, 1992.

Strewart, W.E., Caracotsios, M., and Sørensen, J.P., Parameter Estimation from Multiresponse Data, AlChE J., 38(5), 641 1992.

Topsøe, H., Clausen, B.S., and Massoth, F.E., Hydrotreating catalysis, in *Catalysis Science and Technology*, Vol. 11, Anderson, J.R. and Boudart, M., Eds., Springer-Verlag, New York, 1996.

Vanrysselberghe, V. and Froment, G.F., Hydrodesulfurization of dibenothiophene on a CoMo/Al$_2$O$_3$ catalyst: reaction networks and kinetics, *Ind. Eng. Chem. Res.*, 35, 3311, 1996.

Watson, B. A., Klein, M. T., and Harding, R. H., Mechanistic modeling of n-heptane cracking on HZSM-5, *Ind. Eng. Chem. Res.*, 35(5), 1506–1516, 1996.

Watson, B. A., Klein M. T., and Harding, R. H., Catalytic cracking of alkylcyclohexanes: modeling the reaction pathways and mechanisms, *Intern. J. Chem. Kinetics*, 29(7), 545, 1997.

Whitehurst, D.D., Isoda, T., and Mochida, I., Present state of the art and future challenges in the hydrodesulfurization of polyaromatic sulfur compounds, *Adv. Catal.*, 42, 345, 1998.

10 Automated Kinetic Modeling of Gas Oil Hydroprocessing

10.1 INTRODUCTION

In the previous three chapters, we discussed molecule-based kinetic modeling of catalytic naphtha reforming at the pathways level (Chapter 7), catalytic hydrocracking of heavy paraffins at the mechanistic level (Chapter 8), and catalytic hydrotreating of naphtha at the pathways level (Chapter 9). In this chapter, we will extend and apply our modeling strategy to gas oil hydroprocessing at the pathways level.

From the process chemistry and feedstock points of view, the material of the preceding three chapters will benefit the modeling of gas oil hydroprocessing in this chapter. In Chapter 7, the bifunctional naphtha reforming model builder showed the starting building blocks for generating a model with the bifunctional metal and acid chemistry, but only for the naphtha range feedstock. Hydroprocessing also involves bifunctional metal–acid catalysts but uses gas oil as the primary feedstock. Chapter 8 covered paraffin hydrocracking chemistry, including the gas oil range, and can be directly used for reactions of paraffins in hydroprocessing modeling. Also, the hydrodesulfurization (HDS) and hydrodenitrogenation (HDN) chemistry of Chapter 9 can be extended and applied to the gas oil range in the hydroprocessing process. From the model building point of view, the major challenges in gas oil hydroprocessing lie in the much more complex feedstock and thus much more complex reaction networks.

Catalytic hydroprocessing, like reforming and fluid catalytic cracking (FCC), is one of the most important large-scale petroleum refining processes. Its applications include the processing of heavier feedstocks such as gas oil, the production of high performance lubricants, and the introduction of cleaner-burning fuels. In order to model the hydroprocessing process, as with other complex processes, traditional process models have implemented lumped kinetics schemes, where the molecules are grouped by global properties, such as boiling point or solubility. Molecular information is thus obscured due to the multicomponent nature of each lump. However, increasingly stringent environmental regulations have focused attention on the molecular composition of petroleum feedstocks and their refined products. Modern analytical measurements indicate the existence of $O(10^5)$ unique molecules in petroleum feedstocks. Modeling approaches that allow for reaction of complex feeds and prediction of molecular properties thus require an unprecedented level of molecular detail. In a deterministic modeling approach, each species corresponds to a differential equation; therefore, not only the solution but also the formulation of the implied model are formidable. This motivated us to develop computer algorithms to formulate as well as to solve the model, thereby allowing the process chemists and engineers to focus on the basic chemistry and reaction rules.

In this chapter, this automated kinetic modeling approach is applied to gas oil hydroprocessing, with the focus on gas oil hydrocracking, HDS, and HDN chemistry. The remainder of this chapter is presented in the following sections. The first section describes catalytic hydroprocessing chemistry at the reaction pathways level and the modeling approach with the associated reaction matrices and reaction rules. The following section discusses the technical specifications and details of reactions for each compound class. Then the model building process is automated and a gas oil hydroprocessing model is presented and validated. Finally, we will summarize the results of this work.

10.2 MODELING APPROACH

Prior work in hydrocarbon conversion reactions provides a wealth of kinetic and mechanistic information for modeling the hydroprocessing process on the molecular level (Coonradt et al., 1964; Korre, 1995; Russell, 1996). This information can be combined with graph-theoretic capabilities (Broadbelt et al., 1994a,b 1996), explained in Chapter 2, to build candidate molecular models on the computer.

As discussed in Chapter 3, much of the complexity is statistical or combinatorial, and the large demand for reactions and rate parameters in the molecule-based kinetic models can be handled by organizing the reactions that involve similar transition states into one reaction family, the kinetics of which are constrained to follow the linear free energy relationship (LFER) or quantitative structure-reactivity correlations (QSRC).

Applying these ideas to gas oil hydroprocessing starts with an examination of its feedstock. The gas oil feedstock for the hydroprocessing process contains molecules that can be grouped into aromatics, hydroaromatics, naphthenes, paraffins, olefins, sulfur, and nitrogen-containing compounds. The main difference between the naphtha (as in Chapter 7) and gas oil feedstocks is the inclusion of multiring aromatic, hydroaromatic, and naphthenic compounds in the latter. These multiring compounds introduce many more sites for reactions by the bifunctional metal–acid chemistry.

The reactions in the hydroprocessing process can be classified into reaction families including ring saturation, ring isomerization, ring opening, ring closure, ring dealkylation, side chain cracking, paraffin isomerization, paraffin cracking, olefin hydrogenation, desulfurization, sulfur saturation, denitrogenation, and nitrogen saturation (Girgis and Gate, 1991; Korre, 1995; Russell, 1996; Hou et al., 1997; Topsøe et al., 1996; Whitehurst et al., 1998). Figure 10.1 lists the example reactions for all the reaction families in the hydroprocessing chemistry.

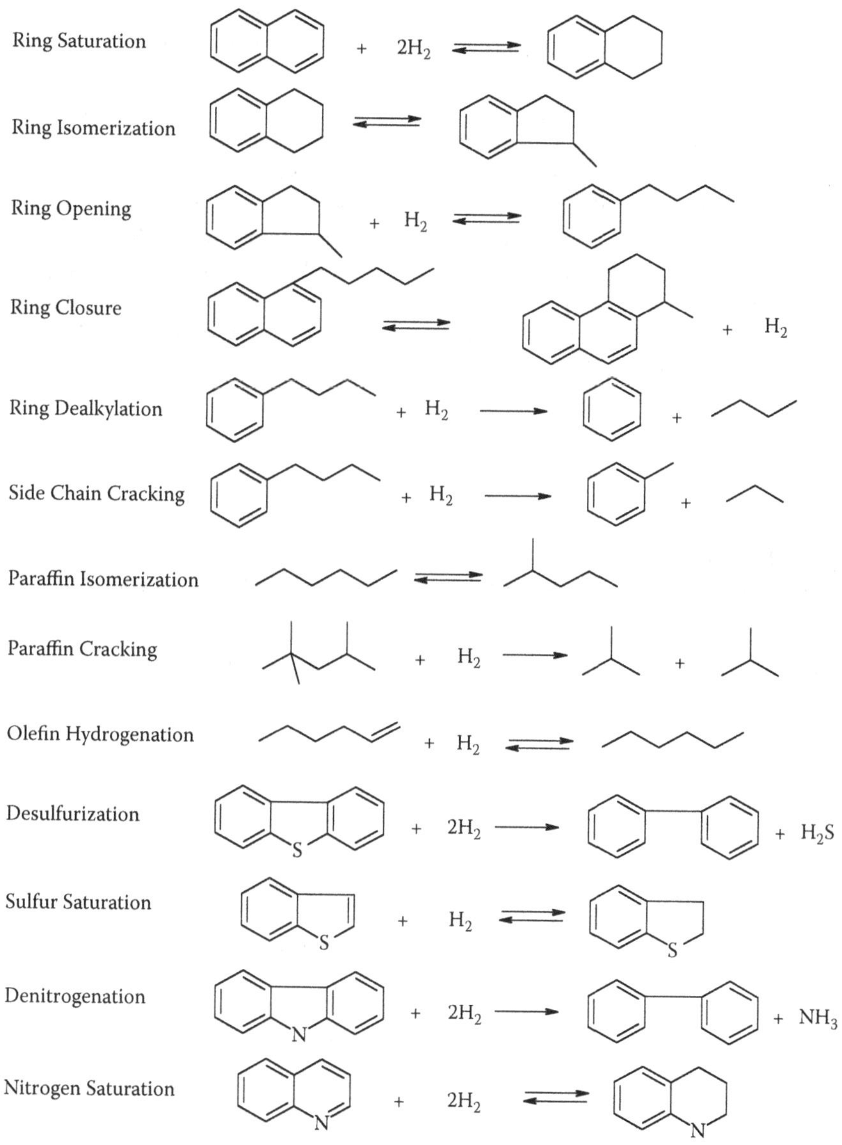

FIGURE 10.1 Reaction families of hydroprocessing at the pathways level.

The molecules in this figure are not representative of the gas oil fraction, which we will discuss in the next section. Also, to clarify the terms, in this work we will be using *saturation* to imply the saturation (hydrogenation) on the ring and *hydrogenation* for others such as olefins.

Thus, the complexity of both the feedstock and the reaction network can be represented by 13 reaction families associated with the reaction matrices, operating on 7 classes of compounds. As a result, a relatively small number of reaction matrices can be used to generate the reaction network with hundreds and thousands of reactions.

The division of all the hydroprocessing reactions into a small number of reaction families, each with an associated transformation operator, can be automated in order to exploit the repetitive nature of the operations. The use of the computer algorithm in the formulation of the model is very important and allows researchers to focus on the basic chemistry and reaction rules of the model. In the next sections, graph-theoretic concepts are exploited to automate hydroprocessing reaction network generation. The reactions pathways in gas oil hydroprocessing and the corresponding reaction matrices for all reaction families are summarized in Figure 10.1 and Table 10.1, respectively. We will discuss the details in the next section.

TABLE 10.1
Reactions Matrices for Reaction Families of Gas Oil Hydroprocessing at the Pathways Level

Reaction Family	Reaction Matrix

Ring saturation (6H case) or nitrogen (N6) saturation (6H case)

Test: an isolated aromatic or six-membered nitrogen ring is required

```
C   0 -1  0  0  0  0  1  0  0  0  0
C  -1  0  0  0  0  0  0  1  0  0  0
C   0  0  0 -1  0  0  0  0  1  0  0
C   0  0 -1  0  0  0  0  0  0  1  0
C   0  0  0  0  0 -1  0  0  0  0  1
C   0  0  0  0 -1  0  0  0  0  0  0
H   1  0  0  0  0  0  0 -1  0  0  0
H   0  1  0  0  0  0 -1  0  0  0  0
H   0  0  1  0  0  0  0  0  0 -1  0
H   0  0  0  1  0  0  0  0 -1  0  0
H   0  0  0  0  1  0  0  0  0  0  0
H   0  0  0  0  0  1  0  0  0  0 -1
```

Ring saturation (4H case) or nitrogen (N6) saturation (4H case)

Test: a terminal aromatic or six-membered nitrogen ring is required

```
C   0 -1  0  0  1  0  0  0
C  -1  0  0  0  0  1  0  0
C   0  0  0 -1  0  0  1  0
C   0  0 -1  0  0  0  0  1
H   1  0  0  0  0 -1  0  0
H   0  1  0  0 -1  0  0  0
H   0  0  1  0  0  0  0 -1
H   0  0  0  1  0  0 -1  0
```

TABLE 10.1 (Continued)
Reactions Matrices for Reaction Families of Gas Oil Hydroprocessing at the Pathways Level

Reaction Family	Reaction Matrix

Ring saturation (2H case 1) or
nitrogen (N6) saturation (2H case 1)

Test: a middle aromatic or six-
membered nitrogen ring is required

C	0	-1	0	0	1	0
C	-1	0	1	0	0	0
C	0	1	0	-1	0	0
C	0	0	-1	0	0	1
H	1	0	0	0	0	-1
H	0	0	0	1	-1	0

Ring saturation (2H case 2) or
nitrogen (N6) saturation (2H case 2)

Test: a middle aromatic or six-
membered nitrogen ring is required

C	0	-1	1	0
C	-1	0	0	1
H	1	0	0	-1
H	0	1	-1	0

Ring isomerization or paraffin
isomerization

Test: the string C—C—C—H is
required

C	0	-1	1	0
C	-1	0	0	1
C	1	0	0	-1
H	0	1	-1	0

Ring opening or ring dealkylation or
side chain cracking or paraffin
cracking

Test: the string C—C is required

C	0	-1	1	0
C	-1	0	0	1
H	1	0	0	-1
H	0	1	-1	0

Ring closure

Test: the string H—C—C—H is
required

H	0	-1	0	1
C	-1	0	1	0
C	0	1	0	-1
H	1	0	-1	0

Olefin hydrogenation

Test: the string C=C is required

C	0	-1	1	0
C	-1	0	0	1
H	1	0	0	-1
H	0	1	-1	0

(Continued)

TABLE 10.1 (Continued)
Reactions Matrices for Reaction Families of Gas Oil Hydroprocessing at the Pathways Level

Reaction Family	Reaction Matrix

Sulfur saturation (6H case) or nitrogen (N5) saturation (6H case)

Test: an isolated aromatic ring is required

```
C    0 -1  0  0  0  0  0  1  0  0  0  0  0
C   -1  0  0  0  0  0  0  0  1  0  0  0  0
C    0  0  0 -1  0  0  0  0  0  1  0  0  0
C    0  0 -1  0  0  0  0  0  0  0  1  0  0
C    0  0  0  0  0  0 -1  0  0  0  0  1  0
C    0  0  0  0 -1  0  0  0  0  0  0  0  1
H    1  0  0  0  0  0  0  0 -1  0  0  0  0
H    0  1  0  0  0  0 -1  0  0  0  0  0  0
H    0  0  1  0  0  0  0  0  0 -1  0  0
H    0  0  0  1  0  0  0  0 -1  0  0  0
H    0  0  0  0  1  0  0  0  0  0  0 -1
H    0  0  0  0  0  1  0  0  0  0 -1  0
```

Sulfur saturation (4H case) or nitrogen (N5) saturation (4H case)

Test: the string C=C—C=C on sulfur or five-membered nitrogen ring is required

```
C    0 -1  0  0  1  0  0  0
C   -1  0  0  0  0  1  0  0
C    0  0  0 -1  0  0  1  0
C    0  0 -1  0  0  0  0  1
H    1  0  0  0  0 -1  0  0
H    0  1  0  0 -1  0  0  0
H    0  0  1  0  0  0  0 -1
H    0  0  0  1  0  0 -1  0
```

Sulfur saturation (2H case) or nitrogen (N5) saturation (2H case)

Test: the string C=C on sulfur or five-membered nitrogen ring is required

```
C    0 -1  1  0
C   -1  0  0  1
H    1  0  0 -1
H    0  1 -1  0
```

Desulfurization (C—S case)

Test: the string C—S is required

```
C    0 -1  1  0
S   -1  0  0  1
H    1  0  0 -1
H    0  1 -1  0
```

Desulfurization (S—S case)

Test: the string S—S is required

```
S    0 -1  1  0
S   -1  0  0  1
H    1  0  0 -1
H    0  1 -1  0
```

TABLE 10.1 (Continued)
Reactions Matrices for Reaction Families of Gas Oil Hydroprocessing at the Pathways Level

Reaction Family		Reaction Matrix								
Desulfurization (C—S—C case)	C	0	−1	0	1	0	0	0		
	S	−1	0	−1	0	0	1	1		
Test: the string C—S—C is required	C	0	−1	0	0	1	0	0		
	H	1	0	0	0	−1	0	0		
	H	0	0	1	−1	0	0	0		
	H	0	1	0	0	0	0	−1		
	H	0	1	0	0	0	−1	0		
Denitrogenation (C—N case)	C	0	−1	1	0					
Test: the string C—N is required	N	−1	0	0	1					
	H	1	0	0	−1					
	H	0	1	−1	0					
Denitrogenation (C—N—C case)	C	0	−1	0	1	0	0	0		
	N	−1	0	−1	0	0	1	1		
Test: the string C—N—C is required	C	0	−1	0	0	1	0	0		
	H	1	0	0	0	−1	0	0		
	H	0	0	1	−1	0	0	0		
	H	0	1	0	0	0	0	−1		
	H	0	1	0	0	0	−1	0		

In this application of hydroprocessing kinetics at the pathways level, the rate constant information was organized with a QSRC/LFER for each reaction family shown in Equation 10.1,

$$E^* = E_o^* + \alpha \Delta H_{rxn} \tag{10.1}$$

which relates the activation energy of each reaction to its heat of reaction. A single frequency factor was assumed for each reaction family. The QSRCs developed by Korre (1995) and Russell (1996) and summarized in Chapter 4 for catalytic hydrocracking can be used here for hydrocracking reactions of all the polynuclear aromatic and hydroaromatic compounds. The rate parameters developed for heavy paraffin hydrocracking in Chapter 8 can be used here for all the paraffin reactions. The rate parameters developed for HDS in Chapter 9 can be used for all the sulfur compound reactions.

10.3 MODEL DEVELOPMENT

The gas oil hydroprocessing modeling is very complex due to the complexity of the feedstock. First, we will review gas oil characterization and construct a representative set of molecular structures for gas oil. To constrain the reaction network building, various reaction rules have been applied based on literature and experience, which are summarized in Table 10.2. The specifics and rules for the reactions of each compound class in the gas oil hydroprocessing pathways will be discussed in turn. Then, we will briefly review the hydroprocessing kinetics. Finally, we will discuss how to automate the molecule-based kinetic modeling of gas oil hydroprocessing.

10.3.1 FEEDSTOCK CHARACTERIZATION AND CONSTRUCTION

From the boiling point (b.p.) point of view, the 520 to 610°F cut is normally considered as light gas oil, the 610 to 800°F cut as heavy gas oil, and the 800 to 1000°F cut as vacuum gas oil. From the carbon number point of view, gas oil

---

TABLE 10.2
Rules for Modeling of Gas Oil Hydroprocessing at the Pathways Level

Reaction Family	Reaction Rules
Ring saturation	Ring saturation proceeds in a ring-by-ring manner. Only isolated aromatic rings underwent saturation with six hydrogen atoms; terminal aromatic rings underwent saturation with four hydrogen atoms; and middle aromatic rings underwent saturation with two hydrogen atoms.
Ring isomerization	Terminal six-membered naphthene rings underwent ring isomerization to a five-membered naphthene ring.
Ring opening	Ring opening occurred only when the implied intermediate corresponded to a secondary or tertiary carbon atom.
Ring closure	Only for ring branches with chain length ≥4. Accompanied with dealkylation if side chain length was ≥7.
Ring dealkylation	Ring dealkylation could occur when the side chain length was greater than two.
Side chain cracking	Only methyl, ethyl, and butyl side chains formed.
Paraffin isomerization	Either leads to branch or increase the side chain length. A maximum of three branches allowed. Only methyl or ethyl side chains allowed.
Paraffin cracking	Only type A and B are allowed.
Olefin hydrogenation	All double bonds not on the ring required to be hydrogenated.
Desulfurization or denitrogenation	S or N is completely, directly removed from the ring.
Sulfur saturation or nitrogen saturation	Ring saturation proceeds in a ring-by-ring manner.

TABLE 10.3
Estimated Molecular Structures in Gas Oil

Aromatic Ring Number	Naphthenic Ring Number	Side Chain Length
4	0	0–5
3	1	0–6
3	0	0–10
2	1	0–12
2	0	3–17
1	2	1–14
1	1	4–20
1	0	10–24
0	4	0–11
0	3	2–16
0	2	5–23
0	1	10–25
0	0	16–31

contains hydrocarbons, with carbon number normally ranging from 20 to 40. At the gas oil range, it is very difficult to characterize the feedstock on a molecule-by-molecule basis beyond the rough estimate of molecular structures. For example, a simple correlation of boiling point with molecular structure groups could give us a rough estimate of the molecular structures in the gas oil, as shown in Table 10.3.

In Chapter 2, we developed a much more sophisticated stochastic approach that can be used to synthesize a molecular representation of the complex feedstock. The approach converts routinely available analytical information for gas oils, such as molecular weight distribution, SIMDIS, PIONA (Paraffin, Iso-paraffin, Olefin, Naphthene, and Aromatics) cut, NMR, etc., into probability density functions (PDFs) for molecular attributes, such as number of aromatic rings, number of naphthenic rings, side chain length, etc. These PDFs are then sampled to generate molecules and optimized to match the experimental measurements to best represent the gas oil.

Applying this stochastic approach generated the molecular representation of a gas oil with the measured properties described in Table 10.4 (Joshi et al., 1998). The optimized molecular representations and the corresponding mole fractions are shown in Table 10.5. The details of this molecular structure modeling approach and process for complex feedstocks can be found in Chapter 2.

10.3.2 REACTION FAMILIES

The reaction families in hydroprocessing are listed in Section 10.2 and illustrated in Figure 10.1. These reaction families, along with their reaction rules, form the basis of the hydroprocessing model builder. The reaction rules are mainly tailored to constrain the chemistry to the important reactions and keep the model to a

TABLE 10.4
Example Description of Gas Oil Feedstoc K

Property	Value
API gravity, °API	28.4
Molecular weight	358
Hydrogen, wt%	13.2
Distillation, °F	
IBP	401
5 vol%	538
10 vol%	597
20 vol%	661
30 vol%	708
40 vol%	754
50 vol%	795
60 vol%	837
70 vol%	884
80 vol%	936
90 vol%	1005
95 vol%	—
EP	—
Composition, wt%	
Paraffins	26.9
Naphthenes	40.8
Aromatics	32.4

modest size. Table 10.2 lists the rules for each reaction family. In this section, we will review the reactions associated with each compound class.

10.3.2.1 Reactions of Aromatics and Hydroaromatics

The reaction pathways for the hydrocracking model of polynuclear aromatic and hydroaromatic compounds have an extensive literature. Earlier work was performed over slightly acidic Si/Al-supported catalysts (Sullivan et al., 1964; Qader and Hill, 1972; Qader, 1973; Huang et al., 1977; Lemberton and Guisnet, 1984), while more recent publications deal with zeolite catalysts as well (Haynes et al., 1983; Lapinas et al., 1987, 1991; Landau et al., 1992; Korre, 1995; Russell, 1996). Qualitatively, it is generally considered that the main reaction path is serial. Hydrogenation or saturation of an aromatic ring is followed by isomerization of the resulting cyclohexyl moiety to a methyl-cyclopentyl moiety, prior to ring opening to one or more side chains and eventual dealkylation. Figure 10.2 shows the representative reaction pathways for aromatics and hydroaromatics.

Ring saturation of an aromatic nucleus is an essential first step. Saturations generally proceed in a ring-by-ring manner. Saturation of a middle aromatic ring fused to more than one other aromatic ring, where one hydrogen molecule or 2H is

TABLE 10.5
Molecular Representation of a Gas Oil Feedstock

Structure	Mole Fraction	Structure	Mole Fraction
C_5H_{12}	0.0003	C_3H_8	0.0007
	0.001	$C_{10}H_{22}$	0.001
C_8H_{18}	0.002	C_5H_{12}	0.003
$C_{16}H_{34}$	0.009	$C_{14}H_{30}$	0.036
$C_{11}H_{24}$	0.045	$C_{12}H_{26}$	0.005
$C_{10}H_{22}$	0.021	C_7H_{16}	0.028
$C_{17}H_{36}$	0.005	$C_{15}H_{32}$	0.010
$C_{12}H_{26}$	0.028	$C_{22}H_{46}$	0.010
$C_{20}H_{42}$	0.043	$C_{17}H_{35}$	0.057
$C_{13}H_{27}$	0.001	$C_{11}H_{23}$	0.005
C_8H_{17}	0.0007	$C_{18}H_{37}$	0.0078

(Continued)

TABLE 10.5(Continued)
Molecular Representation of a Gas Oil Feedstock

Structure	Mole Fraction	Structure	Mole Fraction
$C_{16}H_{33}$	0.0326	$C_{13}H_{27}$	0.0435
$C_{23}H_{47}$	0.025	$C_{21}H_{43}$	0.106
$C_{18}H_{37}$	0.158	$C_{30}H_{62}$	0.058
$C_{28}H_{58}$	0.036	$C_{27}H_{56}$	0.03
$C_{26}H_{54}$	0.035	$C_{25}H_{52}$	0.056
$C_{23}H_{48}$	0.052	$C_{22}H_{46}$	0.047

added, proceeds relatively promptly; however, this reaction is equilibrium limited, since the resulting hydroaromatic structure has only two sp³ atoms per ring and is sterically strained. Further hydrogenations of the resulting structures were shown to proceed with low rates (Girgis, 1988; Girgis and Gates, 1991), while limited middle

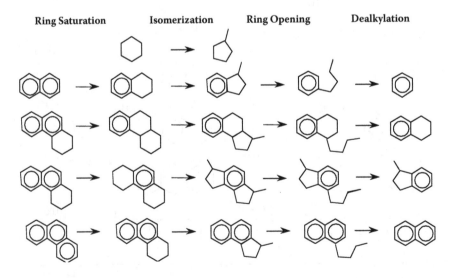

FIGURE 10.2 Reactions of aromatics and hydroaromatics in gas oil hydroprocessing.

ring cleavage reactions to alkyl biphenyls, and eventually alkylbenzenes, were reported (Haynes et al., 1983; Lemberton and Guisnet, 1984; Lapinas, 1989). This pathway appears to be a dead end in the reaction network and is of little significance to conversion hydrocracking. Larger conversions are obtained through pathways involving hydrogenations of terminal aromatic rings (naphthalenic), where two hydrogen molecules or 4H are added. The presence of four sp^3 carbon atoms in the six-membered ring results in many possible acid center transformations. The isolated aromatic rings can also be completely saturated with three hydrogen molecules or 6H under hydrocracking conditions, although this saturation is slow.

Isomerization of a six-membered naphthenic ring to a methyl-substituted five-membered ring occurs prior to ring opening to a butyl side chain. It was suggested that this protonated cyclopropane (PCP)–mediated isomerization is essential to the formation of a more stable, secondary, or tertiary carbenium ion upon ring opening (Sullivan et al., 1964; Qader and Hill, 1972; Haynes et al., 1983; Lemberton and Guisnet, 1984). Thus, ring opening normally occurs only when the implied intermediate corresponds to a secondary or tertiary carbon atom.

Dealkylation of the alkyl-substituted aromatics and hydroaromatics is the last step in the sequence. This was shown to occur with rates proportional to the carbon number of the leaving alkane and was correlated with the heat of formation of the leaving alkyl carbenium ion (Mochida and Yoneda, 1967; Landau, 1991; Landau et al., 1992). Consequently, complete dealkylation is favored, and bare-ring aromatics and hydroaromatics with one ring less than the reactant constitute a large fraction of the final product of the main hydro-cracking pathway. These compounds are subject to the same reaction sequence, namely, ring saturation, ring isomerization, ring opening, and ring dealkylation. Although complete dealkylation is favored, the side chain cracking could happen too, especially for the long alkyl chain. However, based on experimental product spectra of alkylbenzenes and alkyl–polynuclear aromatics (PNAs), only methyl, ethyl, and butyl side chains were observed after the side chain cracking (Russell, 1996).

When the alkyl side chain on the aromatics or hydroaromatics is equal or longer than four, ring closure reaction can happen when the first four carbons on the side chain are closed as a naphthenic ring. When the alkyl side chain on the alkylbenzene is longer than seven, the ring closure reaction is often accompanied by a simultaneous dealkylation (Russell, 1996). This seems to occur less frequently for alkyl-PNAs. With the ring closure reaction pathway, increasingly more rings can build on top of the PNA core. Thus, from the kinetic modeling point of view, the ring closure reaction may serve as the reaction pathway to build up the coke precursor or form coke.

Finally, the recursive nature of PNA hydrocracking pathways suggests the development of QSRCs for all reaction families, ring saturation, ring isomerization, ring opening, and ring dealkylation reactions. As discussed in Chapter 4, rate, equilibrium, and adsorption parameters within each family can be calculated as a function of a reactivity index, pertinent to each reaction, and a small number of regressed parameters. All the developed QSRCs for PNA and alkyl-PNA

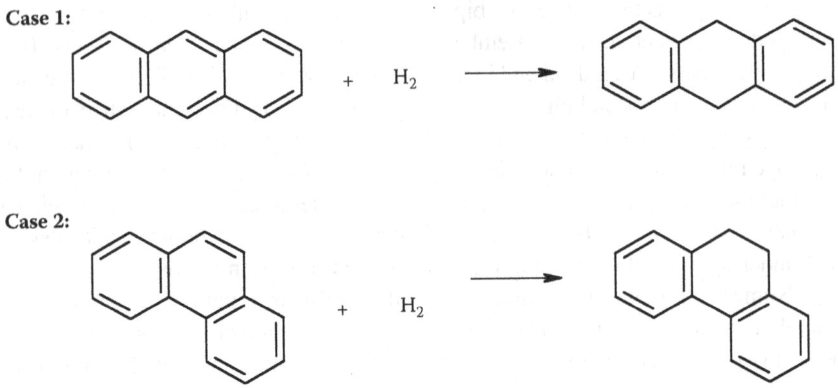

Case 1:

Case 2:

FIGURE 10.3 Two kinds of reactions for the ring saturation (2H case) of a middle aromatic ring.

hydrocracking by Korre (1995) and Russell (1996) were summarized in Chapter 4 and will be used to solve the final kinetic model.

The reaction matrices for all the above-mentioned reaction families are summarized in Table 10.1 and the reaction rules are summarized in Table 10.2. Only the ring saturation 2H case for the middle aromatic ring needs some clarification. The two existing cases for saturation of the middle aromatic ring require two different reaction matrices — one for anthracene (case 1) and one for phenanthrene situation (case 2), as shown in Figure 10.3.

10.3.2.2 Reactions of Naphthenes

Naphthene hydrocracking has been studied mostly for the small naphthenic model compounds (Langlois and Sullivan, 1970; Lazzaraga et al., 1973; Beelen et al., 1973). These studies have shown that the main reactions of single ring naphthenes on bifunctional hydrocracking catalysts are ring isomerization and ring opening. From the mechanistic point of view, the most important difference between paraffinic and naphthenic carbenium ions is the difficulty in cracking the naphthenic ring. A special kind of reaction for naphthenes, called the paring reaction, discovered by a group at Chevron in the early 1960s (Langlois and Sullivan, 1970), alkylates cyclohexanes, with a total of 10 to 12 carbon atoms and hydrocracks in a highly selective manner; the alkyl groups are "pared" from the feed molecule, and the principle products are isobutane and a ring product with four carbon atoms less than the original naphthene. The reaction mechanism can be explained by the carbenium chemistry where the driving route is the A-type β-scission (dealkylation) after extensive skeletal isomerizations (Weitkamp et al., 1984).

Very few studies have been reported on hydrocracking of multiring naphthenes. The main reaction pathways, similar to multiring hydroaromatics, include the reaction sequence of ring isomerization, ring opening, and ring dealkylation.

The reaction matrices are summarized in Table 10.1, and the reaction rules are summarized in Table 10.2.

10.3.2.3 Reactions of Paraffins

We extensively discussed the reaction mechanism and reaction pathways of paraffin hydrocracking in Chapter 8. Extensive skeletal isomerizations precede the cracking reactions; PCP isomerization always leads to branching; A-type cracking always leads to branched isomers; B-type cracking always leads to normal or branched isomers; all the cracking products normally come from A-type cracking of tri-branched isomers or B-type cracking of di-branched isomers.

Here we apply these findings to paraffin hydrocracking as part of the gas oil hydroprocessing. To rationalize the product spectrum and the model size, we have also applied the following reaction rules: paraffin isomerization only leads to methyl or ethyl branches, and a maximum of three branches was allowed; only the type-A cracking and type-B cracking of isoparaffins were allowed; and methane and ethane were not produced after paraffin cracking because of the implied intermediate unstable primary ions.

Table 10.1 and 10.2 summarize the reaction matrices and reaction rules for both the paraffin isomerization and paraffin cracking.

10.3.2.4 Reactions of Olefins

Very low concentrations of olefins may exist in the gas oil feedstock, and some double bonds may be produced in some reaction pathways; for example, the desulfurization of benzothiophene may produce styrene. However, due to the high pressure of hydrogen in the hydroprocessing process, there is no need to model olefins explicitly since their saturation is very fast kinetics and is hydrogenated immediately even if it exists. For the olefins already in the feed or the special cases as we mentioned above, they are hydrogenated directly.

10.3.2.5 Reactions of Sulfur Compounds

In Chapter 9, we discussed the various sulfur compounds and their HDS chemistry. There also exist the following sulfur compound types in gas oil: mercaptans (or thiols), sulfides, disulfides, thiophenes (T), benzothiophenes (BT), dibenzothiophenes (DBT), and their alkyl and hydrogenated derivatives, as illustrated in Figure 10.4.

The mercaptans, sulfides, and disulfides have fast kinetics and can be easily desulfurized. The T, BT, DBT, and their alkyl derivatives can go through either the sulfur saturation reaction or the direct desulfurization reaction, as illustrated in Figure 9.3. T, BT, and DBT are increasingly difficult to remove; alkyl-DBTs, especially those with alkyl chains on the 4 and 6 positions, are the most difficult to remove. To model the HDS chemistry rigorously, it is necessary to incorporate all the representative molecular structures with substituents at both significant (α to sulfur) and nonsignificant positions. The structural approximation concept introduced in

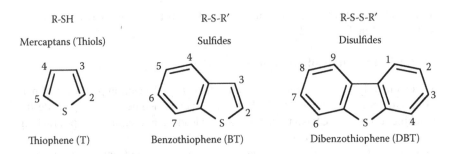

FIGURE 10.4 Representative molecular types of sulfur compounds.

Chapter 9, which accounts for the steric and electronic effects of substituents to the Kinetic reaction rates of thiophenic compounds of the kinetic reaction rates, can also be used to accurately model the sulfur reaction kinetics in gas oil hydroprocessing.

The corresponding reaction matrices for various sulfur saturation and desulfurization reactions are listed in Table 10.1. No specific reaction rules are required for sulfur saturations, except that it proceeds in a ring-by-ring manner; in direct desulfurization, the sulfur is directly removed completely from the ring.

10.3.2.6   Reactions of Nitrogen Compounds

As we briefly discussed in Chapter 9, the nitrogen compounds that would normally be found in the gas oil hydroprocessing feed can be classified into three categories as illustrated in Figure 10.5: basic nitrogen compounds, which are generally associated with a six-member ring, such as pyridine and quinoline; nonbasic nitrogen compounds, which are generally associated with a five-member ring as in indole or carbazole; and the others such as anilines and amines, which are not common in crude oils but may be present in the HDN reaction network of the above nitrogen compounds. The complexity of the nitrogen compounds makes denitrogenation even more difficult than desulfurization.

The HDN chemistry is not as well studied and understood as the HDS chemistry. However, from the modeling point of view, it could be treated like the HDS reaction pathways in the sense that two generic reaction families, nitrogen-compound saturation and direct denitrogenation, could be used to describe the reactions. The amines and anilines, like mercaptans and sulfides, can be easily denitrogenated. As we can see from Figure 10.5, the molecular structures for the nonbasic five-membered nitrogen rings (N5) (such as pyrole, indole, and carbazole) are just like the sulfur rings (such as T, BT, and DBT, respectively). Therefore, their reaction pathways are similar. However, all the basic six-membered nitrogen rings (N6), from the molecular structure point of view, can be treated like aromatic rings. Thus, their reaction pathways also include the various ring saturations and ring opening, except that the saturated six-membered nitrogen ring can be opened without isomerization. This allows direct nitrogen elimination. H_2S has a promotional effect in the C–N bond breaking (denitrogenation)

Basic: 6-Member Rings (N6)

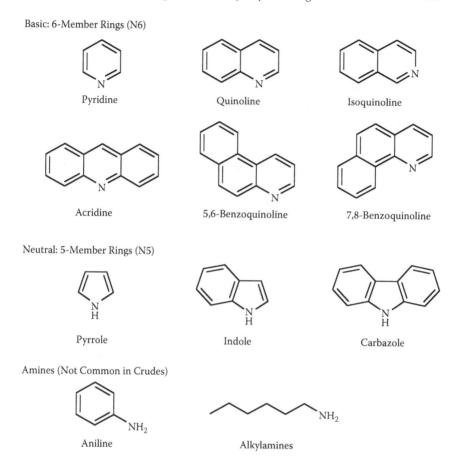

FIGURE 10.5 Representative molecular types of nitrogen compounds.

and can be considered through the rate law describing HDN kinetics (Massoth et al., 1990).

The reaction matrices of nitrogen saturations and direct denitrogenation for N5 are like those for thiophenic compounds; those for N6 are like those for aromatic rings, which are summarized in Table 10.1. The reaction rules are summarized in Table 10.2.

10.3.3 KINETICS: LHHW FORMALISM

Many kinetic studies have been carried out, and kinetic models have been developed for hydrocracking and hydroprocessing reactions of model compounds. A few studies have been published regarding the lumped kinetics of hydrocracking of gas oil and vacuum distillate (Qader and Hill, 1969; Stangeland, 1974; Laxminarasimhan et al., 1996).

To extend the LHHW formalism to complex process chemistry such as gas oil hydroprocessing involving hundreds or thousands of components, several things need to be addressed. First, for any reaction in the complex reaction network, the denominator adsorption group should extend to $(1 + \Sigma K_i p_i)$ to take into account the total inhibition effect of all the components on the reactive catalytic site. Also, for the catalytic systems containing different active sites for different reactions, the rate law should be formulated separately for different sites. Korre (1995) has successfully developed the following dual site LHHW rate law for the PNA hydrocracking chemistry to account for the dual function (both metal and acid) of the hydrocracking catalyst. The rate for each product was obtained as a summation of the rates of its transformations on the metal and the acid sites:

$$\frac{dC_i}{dt} = \frac{\sum_j k_{ji}(C_j - C_i / K_{ji})}{D_H} + \frac{\sum_l k_{li}(C_l - C_i / K_{li})}{D_A} \tag{10.2}$$

where, C_i, C_j, and C_l are component concentrations (mol l^{-1}); k_{ji} and k_{li} are combined numerator rate parameters (l kg$_{cat}$$^{-1}$ s^{-1}) (including intrinsic rate, adsorption constant contributions, and hydrogen pressure where applicable); and K_{ji} and K_{li} are the equilibrium ratios (mol$_i$/mol$_j$/P$_{H2}$^{n}). D_H and D_A are the adsorption groups for the metal (hydrogenation, subscript H) and zeolite (acid, subscript A) sites respectively, defined in Equation 10.3:

$$D_H = 1 + \sum_i K_i^H C_i, \quad D_A = 1 + \sum_i K_i^A C_i \tag{10.3}$$

In Equation 10.3, C_i represents the component concentration (mol l^{-1}), and K_i^H and K_i^A (l mol^{-1}) represents individual component adsorption constants on the metal and acid sites, respectively. Implicit assumptions are surface reaction as the rate determining step and the unity adsorption exponent group in all cases. Equation 10.2 and Equation 10.3 form the basis of the LHHW rate law to describe the gas oil hydroprocessing kinetics. Only a couple of modifications need to be made for the HDS and HDN kinetics. For HDS kinetics, the LHHW form is summarized in Equation 9.1, where the exponent of inhibition or adsorption term was taken as 3, and the overall rate adjustment factor was introduced to account for the total steric and electronic effects of substituents on thiophenic compounds compared with their nonsubstituted parent molecule. For HDN kinetics, a similar LHHW form can be utilized, but the adsorption exponent was found to be 2 and the rate on the direct denitrogenation site needed to be multiplied by $[1 + (K_{H_2S}C_s)^{1/2}]$ to take into account the promotional effect of H$_2$S on C–N bond breaking (Massoth et al., 1990; Pille and Froment, 1997).

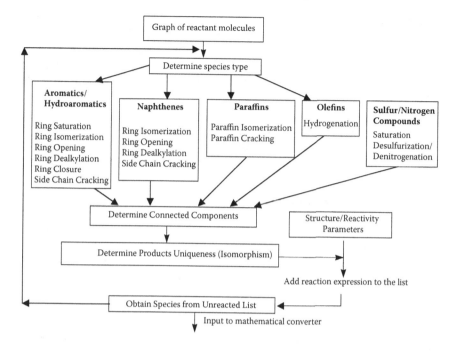

FIGURE 10.6 Algorithm for automated reaction network generation for gas oil hydroprocessing at the pathways level.

10.3.4 AUTOMATED MODEL BUILDING

The reaction matrices of Table 10.1 and the reaction rules of Table 10.2 were used to construct various hydroprocessing kinetic models for the gas oil feedstock.

The model building algorithm used to generate the hydroprocessing models was similar to those for other complex processes such as naphtha reforming, but with the much more enhanced capability of handling multiring compounds in gas oil. Figure 10.6 shows the algorithm for generation of the reaction network. The model builder classifies the graph codes of the molecules into the compound classes of aromatics and hydroaromatics, naphthenes, paraffins, olefins, sulfur compounds, and nitrogen compounds. The reactions allowed by the tests and rules are then carried out by adding the reaction matrix to the reduced permuted reactant matrix. The products are checked for isomorphism (structural uniqueness) and identified as to their compound classes.

The current version of the hydroprocessing model builder includes several enhanced features. Among these is a set of reaction matrices and rules for all types of ring saturation reactions, including two, four, and six hydrogen atoms, and other new pathways reactions for multiring compounds with the flexibility for changing them as new experiments or insights become available. The model builder also includes a library of gas oil molecules. This is composed of information about the atoms and their connectivities, the unique string codes, heats of formations, and IUPAC names. The reactivity indices required for various reactions, such as number of rings, side chain length, and heat of reaction, are calculated "on-the-fly," and the rate expressions are written in a file as an input to the converter. The mathematical converter then generates the rate equations in the LHHW form. The reactor model can then be solved for the case of batch, PFR, or fixed-bed reactor.

The automated model builder of gas oil hydroprocessing provides great flexibility and allows researchers to build complex kinetic models in seconds or minutes and test out various "what-if" scenarios, which would be extremely tedious, if not impossible, to do manually.

10.4 RESULTS AND DISCUSSION

A detailed molecule-based kinetic model for a heavy gas oil hydroprocessing process containing 534 species and 1727 reactions has been developed. The complete reaction model was built automatically in 255 CPU seconds and solves once through in less than 1 minute on an Intel Pentium II 200-MHz machine. The statistics of the model are summarized in Table 10.6.

A molecular representation of a heavy gas oil was constructed, including various representative molecular structures such as pyrene; methylphenylanthracene; chrysene; naphthacene; butylphenanthrene; phenanthrene; anthracenel tetrahydroanthracene; dodecyl-, pentadecyl-, and nonadecylbenzene; nonylnaphthalene; etc.

TABLE 10.6
Statistics of the Gas Oil Hydroprocessing Model

Species	Number	Reactions	Number
Aromatics/hydroaromatics	231	Ring saturation*	454
Naphthenes	93	Ring isomerization*	194
Paraffins	96	Ring opening*	114
Olefins	26	Ring dealkylation	152
Sulfur compounds	74	Side chain cracking	72
Nitrogen compounds	11	Ring closure*	6
H_2	1	Paraffin isomerization*	300
H_2S	1	Paraffin cracking	110
NH_3	1	Olefin hydrogenation*	132
		Sulfur saturation*	100
		Desulfurization	74
		Nitrogen saturation	10
		Denitrogenation	9
Total number of species	534	Total number of reactions	1727

*Reversible reactions in the reaction network.

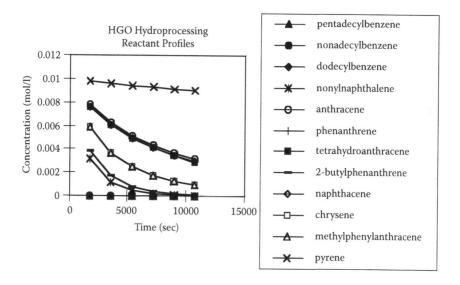

FIGURE 10.7 Reactant conversion profiles in the heavy gas oil (HGO) hydroprocessing model.

The gas oil hydroprocessing reaction network was then built automatically on a computer. A PFR reactor model was solved using experimentally derived QSRCs developed by Korre (1995) and Russell (1996) as documented in Chapter 4. Figure 10.7 shows conversion profiles for some of the reactants. The PNAs were consumed rapidly, except for pyrene, which reacted relatively slowly. Some selected heavy products are shown in Figure 10.8. The PNAs saturated through tetrahydro, octahydro, and finally the perhydro products. Figure 10.8 also shows some lighter products, including the paraffins formed when the side chains of the PNAs dealkylated. Alkylnaphthalenes underwent secondary cracking through ring openings. The model showed good agreement and right trends of the molecular conversions in the process stream with the experimental observations.

10.5 SUMMARY AND CONCLUSIONS

The enhanced automated version of detailed kinetic modeling capability was applied to the gas oil hydroprocessing process with all the process chemistry modules fully integrated together, including hydrocracking chemistry for aromatics, hydroaromatics, naphthenes, paraffins, and olefins; HDS chemistry for sulfur-containing compounds; and HDN chemistry for nitrogen-containing compounds. The major enhancements in the model building tools lie in the capability to handle various multiring aromatic, hydroaromatic, naphthenic, and heterocyclic compounds and the new reaction families, including ring saturation, ring isomerization, ring opening, ring dealkylation, ring closure, side chain cracking, sulfur saturation, nitrogen saturation, desulfurization, and denitrogenation reactions.

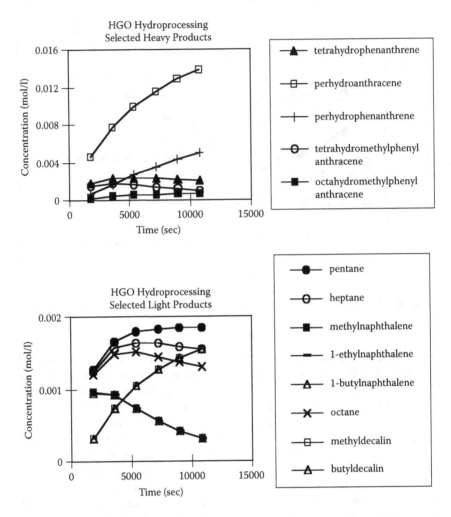

FIGURE 10.8 Product conversion profiles in the heavy gas oil (HGO) hydroprocessing model.

With the increase in the complexity of the molecular structures of multiring compounds, their reaction pathways are also getting more complex. For example, there are three cases in the ring saturation reaction alone: the 6H case for saturation of an isolated aromatic ring, the 4H case for a terminal ring, and the 2H case for the middle ring. Furthermore, the 2H case needs to be handled separately by two different reaction matrices for both the anthracene and phenanthrene cases.

A comprehensive molecule-based heavy gas oil hydroprocessing kinetic model with 534 species and 1727 reactions has been developed and can be easily rebuilt and customized with new experimental data. The model shows good agreement and right trends of the molecular conversions in the process stream with the experimental observations and can be further tuned and optimized with more available data.

This automated kinetic modeling capability for gas oil hydroprocessing has shifted the modeling strategy of this complex process. It provides a great flexibility that allows the users to build various candidate models fast, test out various scenarios, optimize the model adaptively, and select the best model. This development process also demonstrates that the developed generic kinetic modeling tools can be extended to handle various complex reaction systems and significantly speed up the kinetic model development process.

REFERENCES

Beelen, J.M., Ponec, V., and Sachtler, W.M.H., Reactions of cyclopropane on nickel and nickel-copper alloys, *J. Catal.*, 28(3), 376, 1973.

Broadbelt, L.J., Stark, S.M., and Klein, M.T., Computer generated pyrolysis modeling: on-the-fly generation of species, reactions, and rates, *I. & E. C. Res.*, 33(4), 790–799, 1994a.

Broadbelt, L.J., Stark, S.M., and Klein, M.T., Computer generated reaction networks: on-the-fly calculation of species properties using computational quantum chemistry, *Chem. Eng. Sci.*, 49, 4991–5101, 1994b.

Broadbelt, L.J., Stark, S.M., and Klein, M.T., Computer generated reaction modeling: decomposition and encoding algorithms for determining species uniqueness, *Comput. Chem. Eng.*, 20(2), 113–129, 1996.

Campbell, D.M. and Klein, M.T., Construction of a molecular representation of a complex feedstock by Monte Carlo and quadrature methods, *Appl. Catal. AGEN*, 160, 41–54, 1997.

Coonradt, H.L. and Garwood, W.E., Mechanism of hydrocracking reactions of paraffins and olefins, *I. & E. C. Proc. Des. Dev.*, 3, 38, 1964.

Girgis, M., Reaction Networks, Kinetics and Inhibition in the Hydroprocessing of Simulated Heavy Coal Liquids, Ph.D. thesis, University of Delaware, Newark, 1988.

Girgis, M.J. and Gates, B.C., Reactivities, reaction networks, and kinetics in high-pressure catalytic hydroprocessing, *Ind. Eng. Chem. Res.*, 30(9), 2021–2058, 1991.

Haynes, H.W.J., Parcher, J.F., and Helmer, N.E., Hydrocracking polycyclic hydrocarbons over a dual-functional zeolite (faujacite)-based catalyst, *Ind. Eng. Chem. Process Des. Dev.*, 22, 401–409, 1983.

Hou G., Mizan T.I., and Klein M.T., Computer-Assisted Kinetic Modeling of Hydroprocessing, Symposium on Catalysis in Fuel Processing and Environmental Protection, ACS Preprint, 1997.

Huang, C.-S., Wang, K.-C., and Haynes, H.W.J., Hydrogenation of phenanthrene over a commercial cobalt molybdenum sulfide catalyst under severe reaction conditions, In *Liquid Fuels from Coal*, Academic Press, New York, 1977, pp. 63–78.

Joshi, P.V., Iyer, S.D., Klein, M.T., automated Kinetic modeling of gas oil catalytic cracking, *Rev. Process Chem. Engin.*, 1, 111–140, 1998.

Korre, S.C., Quantitative Structure/Reactivity Correlations as a Reaction Engineering Tool: Applications to Hydrocracking of Polynuclear Aromatics, Ph.D. thesis, University of Delaware, 1995.

Landau, R.N., *Chemical Modeling of the Hydroprocessing of Heavy Oil Feedstocks*, Ph.D. thesis, University of Delaware, Newark, 1991.

Landau, R.N., Korre, S.C., Neurock, M.N., Klein, M.T., and Quann, R.J., Hydrocracking of heavy oils: development of structure/reactivity correlations for kinetics, *Am. Chem. Soc. Div. Fuel Chem. Prepr.,* 37(4), 1871, 1992.

Langlois, G.E. and Sullivan, R.F., Chemisty of hydrocracking, *Adv. Chem. Ser. 97,* 38, 1970.

Lapinas, A.T., Catalytic Hydrocracking of Fused-Ring Aromatic Compounds: Chemical Reaction Pathways, Kinetics and Mechanisms, Ph.D. thesis, University of Delaware, Newark 1989.

Lapinas, A.T., Klein, M.T., Gates, B.C., Macris, A., and Lyons, J.E., Catalytic hydrogenation and hydrocracking of fluoranthene: reaction pathways and kinetics, *Ind. Eng. Chem. Res.,* 26, 1026–1033, 1987.

Lapinas, A.T., Klein, M.T., Gates, B.C., Macris, A., and Lyons, J.E., Catalytic hydrogenation and hydrocracking of fluorene: reaction pathways, kinetics and mechanisms, *Ind. Eng. Chem Res.,* 30, 42–50, 1991.

Laxminarasimhan, C.S., Verma, R.P., and Ramachandran, P.A., Continuous lumping model for simulation of hydrocracking, *AIChE J.,* 42(9), 2645–2653, 1996.

Lemberton, J.-L. and Guisnet, M., Phenanthrene hydroconversion as a potential test reaction for the hydrogenating and cracking properties of coal hydroliquefaction catalysts, *Appl. Catal.,* 13, 181–192, 1984.

Lazzaraga, M.G. and Voorhies, A., Jr., Hydroisomerization and Hydrocracking of Cyclohexane in the Presence of a Palladium-Hydrogen-Faujasite Catalyst, *Ind. Eng. Chem. Prod. Res. Dev.,* 12, 194, 1973.

Massoth, F.E., Balushami, K., Shabtai, J., Catalytic functionalities of supported sulfides. VI. The effect of H_2S promotion on the kinetics of indole hydrogenolysis, *J. Catal.,* 122, 256, 1990.

Mochida, I. and Yoneda, Y., Linear free energy relationships in heterogeneous catalysis. I. Dealkylation of alkylbenzenes on cracking catalysts, *J. Catal.,* 7, 386–392, 1967.

Pille, R. and Froment, G., Kinetic study of the hydrodenitrogenation of pyridine and piperidine on a NiMo catalyst, In *Hydrotreatment and Hydrocracking of Oil Fractions,* Groment, G., Delmon, B., and Grange, P., Eds., Elsevier, Amsterdam, 1997, pp. 403–413.

Qader, S.A., Hydrocracking of polynuclear aromatic hydrocarbons over silica-alumina based dual functional catalysts, *J. Inst. Pet.,* 59, 178–187, 1973.

Qader, S.A. and Hill, G.R., Hydrocracking of gas oil, *I&EC Proc. Des. Dev.,* 8(1), 98, 1969.

Qader, S.A. and Hill, G.R., Development of catalysts for the hydrocracking of polynuclear aromatic hydrocarbons, *Am. Chem. Soc. Div. Fuel Chem. Prepr.,* 16, 93–106, 1972.

Scherzer, J. and Gruia, A.J., *Hydrocracking Science and Technology,* Marcel Dekker, New York, 1996.

Stangeland, B.E., A kinetic model for the prediction of hydrocracker yields, *I&EC Proc. Des. Dev.,* 13, 71, 1974.

Sullivan, R.F., Egan, C.J., and Langlois, G.E., Hydrocracking of alkylbenzenes and polycyclic aromatic hydrocarbons on acidic catalysts: evidence for cyclization of the side chains, *J. Catal.,* 3, 183–195, 1964.

Topsøe, H., Clausen, B.S., and Massoth, F.E., In *Catalysis Science and Technology,* Vol. 11, Anderson, J. R. and Boudart, M., Eds., Springer-Verlag, New York, 1996.

Weitkamp, J., Ernst, S., and Karge, H.G., Peculiarities in the conversion of naphthenes on bifunctional catalysts, *Erdoel Kohle-Erdgas-Petreochem.,* 37(10), 457, 1984.

Whitehurst, D.D., Isoda, T., and Mochida, I., *Adv. Catal.,* 42, 345, 1998.

11 Molecular Modeling of Fluid Catalytic Cracking

11.1 INTRODUCTION

Fluid catalytic cracking (FCC) is a major refinery process designed to upgrade heavy petroleum fractions in the carbon range $C14 \leq CN \leq C40$. Typical feedstocks for the FCC process range from light gas oil (LGO) to heavy hydrotreated resids. FCC units employ a high-activity zeolite-based acid catalyst (zeolite Y or USY). The riser reactor is the most common FCC reactor.

Reaction models are useful tools for process design and control of FCC units. They are used routinely for process optimization: determination of appropriate process conditions and selection of feedstock. Traditionally, due to the complexity of the feedstock and associated analytical and computational hurdles, lumped modeling schemes have been employed for FCC reaction models (Weekman and Nace, 1970; John and Wojciechowski, 1975; Jacob et al., 1976). In such schemes, the entire reaction mixture is described in terms of lumps, defined based on boiling point, solubility, or some such physical property. Each lump represents many different molecules with vastly different reactivities. Consequently, lumped reaction models have little chemical significance and limited predictive capabilities.

Due to economic and environmental considerations, the FCC units have come under close scrutiny regarding their molecular product composition. New questions of unprecedented molecular detail have provided impetus for molecular modeling approaches. The new paradigm in reaction modeling is to track each molecule in the feed and the product through the process and to move toward models with fundamental kinetic information and precise predictive capabilities. These notions of reaction modeling at the molecular level have been developed for heterogeneous catalytic processes (Liguras and Allen 1989a,b; Quann and Jaffe, 1992, 1996; Watson and Klein, 1996; Joshi, 1998).

The goal of detailed molecular modeling is to develop models that have good predictive and extrapolative capabilities over a range of process conditions and feedstocks. This requires the incorporation of as much fundamental kinetic information as possible. At the most fundamental level, the development of such reaction models involves modeling of the chemistry at the mechanistic level. Mechanistic models provide several advantages, but their solve time can be very large. In typical industrial applications, such as real time process control, this can be a debilitating limitation. Hence, there is a need to develop techniques that allow faster solution of the model. In the next section, model pruning strategies capable of significantly reducing model solution time will be discussed. Pathways level reaction modeling will then be presented as a relatively CPU-friendly alternative to mechanistic modeling.

11.2 MODEL PRUNING STRATEGIES FOR MECHANISTIC MODELING

11.2.1 MECHANISTIC MODELING

Mechanistic models explicitly account for each molecule and ionic intermediate. Representation of the inherent complexity of FCC chemistry can lead to extremely large reaction networks. The size of the reaction model increases almost exponentially with carbon number (Table 11.1). The use of an automated reaction network builder, as discussed in Chapter 3, makes possible the efficient and inexpensive development of complex reaction networks. However, the mechanistic models are intrinsically stiff and the solution of large models is very CPU intensive.

Clearly, the models need to be pruned at either the model building stage or the model solution stage to reduce the solution time without significantly sacrificing the detailed mechanistic chemistry. This is best done at the model building stage by use of reaction rules to include in the reaction network only reactions that have a significant contribution to process chemistry and product distribution.

11.2.2 RULES BASED REACTION MODELING

11.2.2.1 Reaction Rules

Most model pruning strategies seek to speed up the solution of the model by reducing the size of the model through elimination of reactions without significantly

TABLE 11.1
Size of Various FCC Models as a Function of Cumulative Carbon Number

Cumulative Carbon Number	No. of Model Species	No. of Stable Species	No. of Ions	No. of Reactions	Model Building Time (sec)	Solution Time (sec)
15	225	90	135	3000	1354	400
20	280	110	170	3800	1742	620
25	340	135	205	4700	2286	860
30	425	170	255	5500	2656	1200
35	550	220	330	7000	4240	1840

sacrificing chemical significance. Two characteristics critical to any successful model pruning strategy are:

1. It should be easy to implement. The computational overhead associated with model pruning should be small enough so as not to add significantly to the computational burden.
2. The reduction in the model solution time should not be at the expense of chemical significance. The resultant reduced model should retain the desirable predictive and extrapolative capabilities across a range of process conditions and feedstocks.

The most common and powerful model pruning strategy involves an understanding of the process chemistry. The modeler uses his insights into the process chemistry to include only reactions that are kinetically significant and exclude all reactions that do not have a substantial effect on the product distribution. In the context of the automated computer-based model builder NetGen, this is accomplished through the implementation of reaction rules.

The process begins with the observation that the many reactions in a mechanistic model can be organized into a small set of reaction families, usually of order 10. A reaction family is described as a set of reactions that has similar transition states. The complex gas oil FCC chemistry includes nine reaction families, namely, protonation, deprotonation, protolytic cleavage, hydride transfer, hydride/methyl shift, protonated cyclopropane (PCP) isomerization, ring closure, ring openings, and β-scission. Figure 11.1 summarizes the reaction families that describe the FCC chemistry at the mechanistic level.

For each reaction family, there is a minimum feasibility requirement called a test. It is a necessary but not sufficient condition for a reaction to occur. For example, protolytic cleavage can occur only at a C—C bond. However, not all C—C bonds need necessarily undergo protolytic cleavage. Table 11.2 summarizes the tests for each of the reaction families. The table also shows the unique reaction matrix used by the automated model builder to carry out the reactions in each reaction family.

In the case of gas oil FCC, many reactive sites satisfy the minimum feasibility requirement of the test. To include all such possible reactions would lead to an unmanageably large reaction model. Fortunately, it is not essential to model all the possible reactions, but only the kinetically significant ones. This is accomplished by passing the set of all possible reactions through the "colander" of reaction rules. Reaction rules allow user knowledge, hypotheses, or biases to filter the realized reactions to include only the kinetically significant pathways. For example, protolytic cleavage is not allowed for double bonds and cyclic species because other faster reaction pathways, such as protonation and hydride abstraction, dominate the FCC chemistry at conditions of practical interest. The reaction rules used for a gas oil FCC model builder are summarized in Table 11.3.

In spite of judicious choice of reaction rules, the mechanistic FCC gas oil model can have many species and reactions, so the model is too large to solve in a

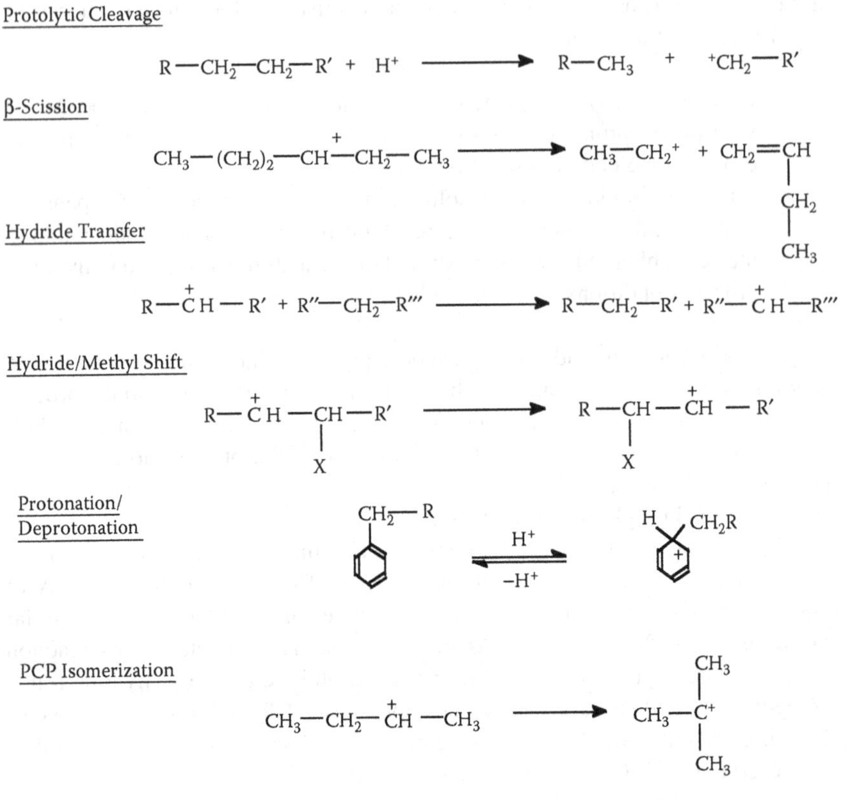

Protolytic Cleavage

$$R-CH_2-CH_2-R' + H^+ \longrightarrow R-CH_3 + {}^+CH_2-R'$$

β-Scission

$$CH_3-(CH_2)_2-\overset{+}{CH}-CH_2-CH_3 \longrightarrow CH_3-CH_2^+ + CH_2{=}CH$$

$$\begin{array}{c} | \\ CH_2 \\ | \\ CH_3 \end{array}$$

Hydride Transfer

$$R-\overset{+}{C}H-R' + R''-CH_2-R''' \longrightarrow R-CH_2-R' + R''-\overset{+}{C}H-R'''$$

Hydride/Methyl Shift

$$R-\overset{+}{C}H-\underset{\underset{X}{|}}{CH}-R' \longrightarrow R-\underset{\underset{X}{|}}{CH}-\overset{+}{C}H-R'$$

Protonation/ Deprotonation

PCP Isomerization

$$CH_3-CH_2-\overset{+}{CH}-CH_3 \longrightarrow CH_3-\underset{\underset{CH_3}{|}}{\overset{\overset{CH_3}{|}}{C^+}}$$

FIGURE 11.1 Mechanistic reaction families for gas oil FCC.

reasonable amount of time. It is useful to realize that the explosion in the number of species and reactions is mostly combinatorial in nature. An effective strategy for reducing the size of the model focuses on curtailing this combinatorial explosion.

11.2.2.2 Stochastic Rules

In FCC chemistry, two reaction families—protolytic cleavage and hydride transfer—lead to a combinatorial explosion of species and reactions. In hydride transfer, each ionic species can abstract H from any of the several C—H bonds in the molecule, resulting in new ionic species. The resultant ionic species undergo further hydride transfer reactions, and so on, thereby setting off the combinatorial explosion. A typical gas oil has molecules with several C—C bonds that are the reactive centers for protolytic cleavage. Each of the many such molecules in a gas oil feedstock can undergo several protolytic cleavages, resulting in many species and reactions. It is not useful to eliminate these reactions through use of reaction rules because each of the reactions is, potentially, a kinetically significant reaction pathway.

TABLE 11.2
Reactions Matrices for Mechanistic Reaction Families of FCC of Gas Oils

Reaction Family   **Reaction Matrix**

Protolytic Cleavage

Test: the string C—C is required

$$R-CH_2-CH_2-R' + H^+ \longrightarrow R-CH_3 + {}^+CH_2-R'$$

$$\begin{array}{c|ccc} C & 0 & -1 & 0 \\ C & -1 & 0 & 1 \\ H+ & 0 & 1 & 0 \end{array}$$

Hydride Transfer

Test: the string C—H is required in a molecule and the string C+ in an ion

$$R-\overset{+}{C}H-R' + R''-CH_2-R''' \longrightarrow R-CH_2-R' + R'-\overset{+}{C}H-R'''$$

$$\begin{array}{c|ccc} C+ & 0 & 1 & 0 \\ C & -1 & 0 & -1 \\ H & 0 & -1 & 0 \end{array}$$

β-Scission

Test: the string C+—C—C is required in the carbenium ion

$$CH_3-(CH_2)_2-\overset{+}{C}H-CH_2-CH_3 \longrightarrow CH_3-CH_2^+ + CH_2=CH$$
$$\overset{\displaystyle |}{CH_2}$$
$$\overset{\displaystyle |}{CH_3}$$

$$\begin{array}{c|ccc} C+ & 0 & 1 & 0 \\ C & 1 & 0 & -1 \\ C & 0 & -1 & 0 \end{array}$$

Methyl Shift/Hydride Shift

Test: the string C+—C—C is required in the carbenium ion

$$R-\overset{+}{C}H-CH-R' \longrightarrow R-CH-\overset{+}{C}H-R'$$
$$\quad\quad | \quad\quad\quad\quad\quad\quad |$$
$$\quad\quad X \quad\quad\quad\quad\quad\quad X$$

$$\begin{array}{c|ccc} C+ & 0 & 0 & 1 \\ C & 0 & 0 & -1 \\ H & 1 & -1 & 0 \end{array}$$

PCP Isomerization

Test: The string C+—C—C is required in the carbenium ion

$$CH_3-CH_2-\overset{+}{C}H-CH_3 \longrightarrow CH_3-\overset{\displaystyle CH^3}{\overset{\displaystyle |}{\underset{\displaystyle |}{C^+}}}$$
$$\quad\quad\quad\quad\quad\quad\quad\quad\quad\quad\quad CH_3$$

$$\begin{array}{c|cccc} C+ & 0 & -1 & 1 & 0 \\ C & -1 & 0 & 0 & 1 \\ C & 1 & 0 & 0 & -1 \\ H & 0 & 1 & -1 & 0 \end{array}$$

Protonation

Test: the string C=C is required

$$\begin{array}{c|ccc} C & 0 & -1 & 1 \\ C & -1 & 0 & 0 \\ H+ & 1 & 0 & 0 \end{array}$$

(Continued)

TABLE 11.2 (Continued)
Reactions Matrices for Mechanistic Reaction Families of FCC of Gas Oils

Reaction Family	Reaction Matrix

Deprotonation

Test: the string C+ —C is required.

$$\begin{array}{c|ccc} C+ & 0 & 0 & 1 \\ C & 0 & 0 & -1 \\ H & 1 & -1 & 0 \end{array}$$

A path forward toward curtailing the combinatorial explosion is through stochastic elimination of protolytic cleavage and hydride transfer reactions. In this scheme, of all the possible kinetically equivalent protolytic cleavage or hydride transfer reactions that a molecule can undergo, only a fraction are allowed. The choice is made through stochastic sampling.

First an exhaustive list of all the possible reactive sites in a molecule is generated. The probability of a reaction at a particular site is defined by a probability density function (PDF). In principle, the probability of reaction at a particular site can account for differences in reaction rate not discernable through quantitative structure reactivity relationships (discussed below). For example, certain large functional groups, due to geometric constraints, may have a lower probability of reacting at particular sites on a zeolite.

Next, random numbers are generated, and the probability density functions sampled to select from the list the specific reactions to be modeled. The reaction rate constants for these reactions are scaled up appropriately to account for only a fraction of all the possible reactions being incorporated into the model. This ensures that the disappearance kinetics for the molecule remains accurate, although only a few of all the possible reactions are allowed.

Consider, for example, the case of protolytic cleavage of linear paraffins. Tetradecane has 13 possible protolytic cleavage sites; pentadecane has 14, hexadecane has 15, and so on. The reaction products are similar in each case, namely, carbenium ions in the range of C_2 to C_{n-1} and alkanes in the range of C_1 of C_{n-2}. Each of the resultant carbenium ions carries out many hydride transfers from all of the linear paraffins and other species, resulting in a combinatorial explosion of the reactions and species. It is possible to select, stochastically, a fraction of the protolytic cleavages for each normal paraffin and a few of the hydride transfers for each carbenium ion. This stochastic elimination of the reactions greatly reduces the number of reactions.

The elimination of the reactions through stochastic sampling does not significantly sacrifice chemical significance or skew the product distribution because of compensation effect. Since several similar molecules undergo similar reactions to give similar products, the scaled up reactions of one molecule compensate for

TABLE 11.3
Rules for Mechanistic Modeling of FCC of Gas Oils

Reaction Family	Reaction Rules
Protolytic Cleavage $R-CH_2-CH_2-R' + H^+ \longrightarrow R-CH_3 + {}^+CH_2-R'$	1. Allowed for all linear paraffins with $C_N > 10$ 2. Not allowed for cyclic species 3. Not allowed for double bonds
Hydride Abstraction $R-\overset{+}{C}H-R' + R'-CH_2-R''' \longrightarrow R-CH_2-R' + R''-\overset{+}{C}H-R'''$	**Donor molecule:** 1. Allowed for all cyclic intermediates with six-membered rings 2. Allowed for allylic and β to branch C atoms for cyclic and isoparaffin species 3. Allowed for multicyclic rings, β to branch on the chain for $C \leq 10$ 4. Not allowed for carbons with single or multiple double bonds **Acceptor ion:** For all noncyclic ions with $C \leq 10$
β-Scission $CH_3-(CH_2)_2-\overset{+}{C}H-CH_2-CH_3 \longrightarrow CH_3-CH_2^+ + CH_2=\underset{\underset{CH_3}{\overset{\mid}{CH_2}}}{\overset{\mid}{CH}}$	1. Only ipso-cracking allowed for aromatics 2. No breakage of diene cyclic species 3. No formation of methylene type compounds 4. No formation of primary carbenium ion for cyclics and acyclics
Hydride Shift/Methyl Shift $R-\underset{X}{\overset{\mid}{C^+H}}-CH-R' \longrightarrow R-\underset{X}{\overset{\mid}{CH}}-C^+H-R'$	1. Only allowed for $C \leq 10$ 2. Not allowed for cyclic species 3. No migration of charge from ring to the side chain 4. Migration to a stable ion, allylic, or branched ions
PCP Isomerization $CH_3-CH_2-\overset{+}{C}H-CH_3 \longrightarrow CH_3-\underset{\underset{CH_3}{\overset{\mid}{C^+}}}{\overset{\overset{\mid}{CH_3}}{}}$	1. Only allowed for $C \leq 10$ 2. Not allowed for cyclic species 3. No formation of primary carbenium ions 4. Formation of a maximum of three branches for a given carbon number

(Continued)

TABLE 11.3 (Continued)
Rules for Mechanistic Modeling of FCC of Gas Oils

Reaction Family	Reaction Rules

Protonation/Deprotonation

1. Allowed for all noncyclic olefins
2. Allowed only at the ipso position for aromatics
3. Primary carbenium ion formation not allowed
4. Not allowed for olefinic carbenium ions having C > 5 and C ≤ 10
5. Not allowed for ions forming $=$ $C=$ (vinyl)-type compounds
6. Not allowed for ions forming methylene-type cyclic species

the eliminated reactions of the other molecules within the same reaction family. Theoretically, the product distribution will be identical only in the limit of the number of reactions, $n \to \infty$. However, the molecules in gas oil are fairly large, and the deviations in product distribution are sufficiently small for gas oil FCC.

To illustrate these concepts, models with stochastic rules were developed for a paraffinic mixture and compared to the model with all reactions explicitly modeled. The feed for the test case was an equimolar mixture of normal paraffins in the carbon range C_{15} to C_{24}. Table 11.4 summarizes the model diagnostics for the various models. Model 1 includes all possible reactions and provides the benchmark for comparison. In model 2 and model 3, only one-half and one-third, respectively, of the possible protolytic cleavages and hydride abstractions were stochastically selected. A simple PDF, zero for the end positions and uniform elsewhere, was used. The equation for the PDF is given by

$$f(n,x) = H(x-2) - H(x-n) \qquad (11.1)$$

where n is the number of carbons, x is the carbon site, and $x \in [1,n]$ H is the heavyside function.

TABLE 11.4
Diagnostics for Various Models

	No. of Species	No. of Reactions	Solution Time (mins)
Model 1	625	10677	216.13
Model 2	417	5022	26.31
Model 3	403	4563	23.53

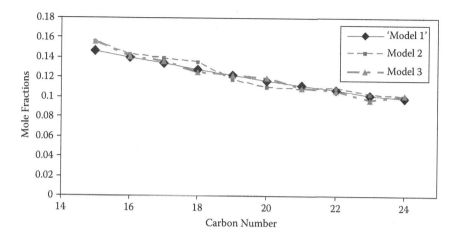

FIGURE 11.2 Paraffin concentrations in product stream for stochastic FCC mechanistic model.

Figure 11.2 compares the results of the three models. The results of all three models are comparable, indicating that it is possible to use stochastic rules without significantly sacrificing chemical significance.

Models 2 and 3 are significantly smaller than model 1 and solve almost eight times faster. The advantage in reducing the number of reactions further from half to a third is not significant, either in terms of model size or solution time. It was observed that the reproducibility of the models suffers if the fraction of reactions allowed to react is made too low.

Joshi (Joshi et al., 1998, Joshi, 1998) applied these ideas to a gas oil FCC model. The stochastic rules Joshi used were to allow random sampling of four sites for compounds with $C_N > 10$ in the case of protolytic cleavages and hydride transfer.

11.3 KINETICS

11.3.1 INTRINSIC KINETICS

The quantitative structure reactivity correlation (QSRC) approach was used to describe the intrinsic kinetics. This approach has been developed for heterogeneous reaction processes (Dunn, 1968) involving both metal and acid functions, such as in hydrocracking (Korre, 1994), catalytic cracking (Dumesic et al., 1993; Watson et al., 1996, 1997), and dealkylation and isomerization (Mochida and Yoneda, 1967a,b,c). The organization of the rate constants into QSRCs greatly reduced the number of model parameters. For example, the representation of 5500 rate constants by 27 QSRC parameters represents a 99.5% reduction in the number of kinetic parameters. Dumesic et al. (1993) and Watson et al. (1996, 1997) successfully used this approach for pure-component acid-catalyzed reactions involving carbenium ion chemistry. The rate constants for all the mechanistic reactions were described in terms of a family-specific Arrhenius factor, A, and the QSRC that related the activation energy to the enthalpy change of reaction, as shown in Equation 11.2:

$$EA = E_o + \alpha \,{}^*\Delta H_{rxn} \qquad (11.2)$$

The rate constant kij for a given reaction i within the reaction family j is given by Equation 11.2.

$$kij = A_j \exp(-(E_{o,j} + \alpha_j \Delta H_{rxn,i})/RT) \qquad (11.3)$$

Each reaction i in the reaction family j is represented by three parameters (A, E_o, and α). The value of α is bounded between 0 and 1. Watson et al. (1996, 1997) found that a value of 0.5 sufficed for most of the reaction families. The values for the Arrhenius factors (A) for each reaction family were chosen from the literature (Watson et al., 1996, 1997). Values ranging from 10^{13} (sec^{-1}) to 10^{18} (sec^{-1}) were used for the unimolecular methyl/hydride-shift, deprotonation, and isomerization reactions. Similarly A_j for β-scission was 10^{16} (sec^{-1}). All the bimolecular reactions were constrained such that $10^2 \le A_j(sec^{-1}\,Pa^{1-n}) \ge 10^4$. The E_{oj} values for reverse reactions were constrained by the enthalpy balance shown in Equation 11.4.

$$E_{backword} - E_{forward} = \Delta H_{rxn} \qquad (11.4a)$$

$$E_{oj,forward} = E_{oj,backword} \qquad (11.4b)$$

$$\alpha_{j,forward} = 1 - \alpha_{j,backward} \qquad (11.4c)$$

A catalyst-dependent parameter (Dumesic et al., 1993; Watson et al., 1996, 1997) that characterizes the relative stabilization of the H+ ion to the other carbenium ions was used as an indication of acidity.

11.3.2 Coking Kinetics

An exponential decay function (Froment and Bischoff, 1962) was used to represent the activity drop caused by coking, as shown in Equation 11.5.

$$\Phi_R = \exp(-\alpha_R C_C) \qquad (11.5)$$

The deactivation function, α_R, represents the decay of H+ ion sites in the model and hence the activity drop as a function of coking.

Polynuclear aromatics (PNAs) and olefins were considered to be the main coke precursors, and a molecular reaction for the formation of coke was considered, as shown in Equation 11.5.

$$\text{Aromatics/olefins} \rightarrow \text{coke} + \gamma\text{hydrogen} \qquad (11.6)$$

The stoichiometric coefficient γ was obtained from the C-to-H ratio of the coke. Thus, a constant empirical formula was assumed for the coke on the catalyst. The rate constant of coke formation was assumed to be constant because of the lack of detailed information as well as to keep the kinetics simple.

11.4 MODEL DIAGNOSTICS AND RESULTS

The reaction matrices of Table 11.2, the rules of Table 11.3, and the kinetic correlations of Equation 11.2 to Equation 11.5 were used to construct a stochastic FCC mechanistic kinetic model for the Yanik et al. (1985) gas oil. Details of model development can be found elsewhere (Joshi et al., 1998). The stochastic FCC mechanistic model contained 425 components (194 molecules and 231 surface ionic species) and 5500 reactions. The reaction breakdown included 400 protolytic cleavage reactions, 4400 hydride transfer reactions, 250 protonation and deprotonation reactions, 250 β-scission reactions, 100 hydride shift and methyl shift reactions, and 100 PCP isomerization reactions. The model solution took about 25 CPU minutes on an IBM RISC 6000. The results have been reproduced in Figure 11.3. The parity between the predicted and experimental results is good.

This demonstrates that the stochastic approach to model reduction is an effective strategy in reducing the size of the model without sacrificing chemical significance. It allowed for the reduction of the complexity of the reaction network, by exploiting the repetitive nature of the process chemistry, and facilitated faster solution of the model.

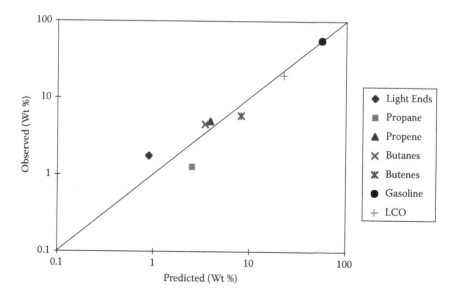

FIGURE 11.3 Parity plot for mechanistic modeling of gas oil FCC.

11.5 MECHANISTIC MODEL LEARNING AS A BASIS FOR PATHWAYS LEVEL MODELING

Development of mechanistic models requires a good understanding of the process chemistry and careful attention to reaction rules. Often the process of developing several candidate models in the search for the optimal model provides useful insights into the process chemistry. Some of the key lessons from the mechanistic modeling of FCC chemistry are as follows.

The product distribution has low sensitivity to changes in the rate parameters for hydride shift, methyl shift, protonation and deprotonation, and isomerization reactions, suggesting that these reactions are fast and essentially equilibrated at typical conditions of industrial operations. The rate determining steps are β-scissions and hydride transfer (Watson et al., 1996).

The larger species crack faster than the smaller species. It was observed that limiting deprotonation reactions for $C_N > 6$ did not significantly affect the product distribution. Joshi (1998) therefore concluded that that β-scission reactions are facile for $C_N > 6$.

Isomerization reactions are limited to species with one to three branches. Joshi (1998) and Watson et al. (1996) found that limiting the branching to three was adequate for explaining the product distribution, and this greatly reduced the size of the reaction network.

The inclusion of coking reactions and modeling catalyst deactivation is required for explaining the observed conversions and product spectra at varying residence time and catalyst loading (Watson et al., 1996).

11.6 PATHWAYS MODELING

Mechanistic models allow incorporation of process chemistry at the most fundamental level, ensuring detailed product distribution and applicability over a wide range of operating conditions and feedstocks. However, the large size and inherent numerical stiffness of such models makes their solution very time consuming. Lumped models can be solved faster but cannot provide detailed molecular information and have limited predictive capabilities. Pathways level models provide a good compromise between speed of solution and detailed molecular product information over a range of operating conditions.

Pathways models are not as complex as mechanistic models because of the exclusion of reaction intermediates, for example, surface species. The exclusion of surface species affords another advantage: it does away almost entirely with the stiffness of the set of ordinary differential equations used to represent the reaction network mathematically. Stiffness in mechanistic models arises because the surface species (carbenium ions) and the molecular species change concentration at vastly different scales along the length of the reactor. In pathways models, all the species (stable molecules) change over relatively longer length scales, and the problem to be solved is essentially a nonstiff system of ordinary differential equations (ODEs). Thus, the absence of stiffness together with the smaller size of the model results in a spectacular speed up in the solution of the pathways level models. The CPU time for solving a pathways model can be a few orders of magnitudes smaller than that for the corresponding mechanistic model.

The fast solution of pathways models allows their use in process optimization and control. They hold the promise of predictive capabilities over a wide range of feedstocks and operating conditions through incorporation of detailed kinetic information and explicit accounting of all important observable molecules. They also offer the opportunity to incorporate various mechanistic insights into the chemistry.

The pathways level model for gas oil FCC developed in the work discussed here included observable molecular species including paraffins, olefins, naphthenes, and aromatics. The model is small enough to be incorporated into the process optimization schemes and large enough to give detailed molecular information.

11.6.1 PATHWAYS MODEL DEVELOPMENT APPROACH

The issues involved in developing a large pathways model are similar to those for mechanistic models. The chemistry must be represented in a rigorous manner. To this end, the reaction network generator, described in Chapter 3, was employed. The number of rate parameters must be kept to a reasonable number to allow tuning of the experimental data. To this end, the concept of reaction families and QSRCs similar to those used for mechanistic model was used.

The application of the reaction network generator depends on the classification of all the catalytic cracking reactions into a small number of reaction families, each with a unique reaction matrix. The catalytic cracking reaction mixture consists of a few observable compound classes (paraffin, isoparaffins, olefins, iso-olefins, naphthenes, and aromatics) that react through a limited number of reaction families (cracking, isomerization, hydrogenation and dehydrogenation, methyl shift, dealkylation, and amortization). Figure 11.4 shows the various reaction families for gas oil catalytic cracking at the pathways level.

The kinetics were described in terms of the Langmuir–Hinshelwood–Hougen–Watson formalism. Due to lack of information about the catalyst, adsorption rate constants were left as optimizable parameters. In principle, these can be calculated from thermodynamic considerations or from a structure reactivity correlation if available (Neurock et al., 1992; Korre, 1994). In general, for an acid catalyst such as that used for catalytic cracking, aromatics adsorb more strongly than naphthenes, which adsorb more strongly than paraffins. Due to lack of experimental information and to limit the number of optimizable parameters, adsorption constants were assumed to be the same for all members of a compound class. Some justification for this comes from preliminary modeling of pure component catalytic cracking. Pure compound studies indicate that the difference in adsorption constants for different members in the same compound class is not important in predicting the product distribution.

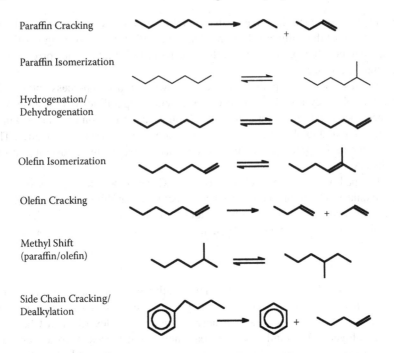

FIGURE 11.4 Reaction families for pathways gas oil catalytic cracking.

11.6.2 PATHWAYS LEVEL REACTION RULES

Mechanistic insights were used to develop and refine the reaction network. These insights were incorporated at the reaction network building stage as reaction rules.

11.6.2.1 Cracking Reactions

All compound classes (paraffins, olefins, naphthenes, and alkyl aromatics) were allowed to undergo cracking reactions. The minimum requirement for cracking reactions is a C—C bond.

Paraffins: All C—C bonds in paraffins were allowed to undergo cracking, with the exception of those that led to the production of methane. This is because mechanistically, methane is formed only through a methyl ion intermediate. All reactions forming methyl ions are energetically unfavorable by about 12 kcal/mol compared to other cracking reactions.

Isoparaffins: Cracking of isoparaffins was allowed only at the branch position and beta to branch. This can be justified by mechanistic considerations.

Energetically, the tertiary ion is more stable than the secondary ion. Hydride abstraction is favored for branch sites, and the resultant ion then undergoes β-scissions. This favored mechanistic sequence of hydride abstraction at the branch site followed by β-scission is modeled at the pathways level as cracking beta to the branch. Another favored pathway at the mechanistic level is the β-scission of the ion with charge at the beta position to the branch to produce the more stable tertiary ion. This suggests that cracking at the branch point is another kinetically dominant pathway for cracking of isoparaffins. The other cracking reactions for isoparaffins not involving branch points are much slower and of relatively minor importance.

Olefins: The favored olefin cracking reactions are beta to the double bond to produce an allylic functional group and gamma to the double bond to produce a conjugated diene. Mechanistically, both these reactions are more facile than other cracking reactions because of the involvement of a more stable allylic ion as an intermediate. For iso-olefins, cracking reactions at the branch and beta to the branch are allowed in addition to the above reactions. However, in no case is cracking at the double bond or adjacent to the double bond allowed.

Side Chain Cracking: For alkyl aromatics, only side chain cleavage is allowed if the alkyl side chain length is greater than 2. This is because protonation at the ipso position followed by β-scission is the dominant pathway.

For alkyl naphthenes, in addition to side chain cleavage, cracking at the beta position on the side chain (side chain length > 2) is allowed. This is justified because in the underlying mechanistic steps, the more stable tertiary ion at the branch point on the ring is an intermediate.

11.6.2.2 Isomerization Reactions

Isomerization reactions are kinetically the fastest reactions in catalytic cracking chemistry. They are equivalent, mechanistically, to PCP isomerization reactions. The driving force for these reactions is the stabilization of the carbenium ion as it forms a tertiary ion from a secondary ion. Isomerization reactions have a dramatic effect on the size of the resultant reaction network. The minimum requirement for isomerization reactions is three adjacent carbons.

At the pathways level, isomerization reactions were allowed for all paraffins and olefins with fewer than 14 carbons. The carbon restriction was useful in keeping the size of the model reasonable. Some justification for this rule comes from knowledge from mechanistic modeling that large molecules crack faster (Watson et al., 1996; Joshi, 1998).

Another rule was to not allow isomerization for molecules with three or more branches. This is based on the knowledge from mechanistic modeling that limiting branching to three is adequate for explaining the product distribution (Watson et al., 1996; Joshi, 1998).

11.6.2.3 Methyl Shift Reactions

Another family of reactions driven by carbenium ion stability is the methyl shift reactions. At the pathways level, the minimum requirement for a methyl shift reaction is a carbon singly bonded to three other carbons. An example of a methyl shift reaction is 2-methyl hexane to 3-methyl hexane. Heats of formation calculations using MOPAC indicate that the 3-methyl–3-hexyl ion is more stable than the 2-methyl–2-hexyl ion by 1.5 kcal/mol. This is because the methyl group stabilizes the carbenium ion. Thus, all mono-branched paraffins and olefins were allowed to undergo a two to three methyl shift. The three to four methyl shift was not allowed because it is energetically neutral and hence not likely a dominant reaction pathway.

11.6.2.4 Hydrogenation and Dehydrogenation Reactions

These reactions are critical in fluid catalytic chemistry because they are responsible for the paraffin-to-olefin ratio observed in the product stream. The minimum requirement for a hydrogenation reaction is a C=C bond, and the minimum requirement for a dehydrogenation reaction is a C—C bond. All hydrogenation and dehydrogenation reactions were modeled as reversible reactions.

All acyclic molecules with fewer than 14 carbons were allowed to undergo these reactions. However, olefins were not allowed to undergo dehydrogenation reactions to form dienes. The carbon number cutoff is justified because larger molecules have a greater tendency to crack.

Through development of various candidate models, it was determined that dehydrogenation to form 1-olefin and 2-olefin need only be modeled to get the required level of detail for the product stream. For isoparaffins, dehydrogenation was allowed only at the branch points.

11.6.2.5 Aromatization

All saturated rings were allowed to undergo this reaction. The only rule imposed was the side chain length be three or less. This is because for large side chain molecules, the side chain cracking reactions are the dominant reactions. This rule is useful in keeping the size of the model reasonable without sacrificing chemical significance.

11.6.3 Coking Kinetics

PNAs and olefins were considered to be the main coke precursors, and a molecular reaction for the formation of coke was considered, as shown in Equation 11.7.

$$\text{Aromatics/olefins} \rightarrow \text{coke} \qquad (11.7)$$

The C-to-H ratio of the coke is usually higher than the aromatics and the olefins, but due to lack of availability of the C-to-H ratio for the coke, the hydrogen production from coking reactions has been neglected. The rate constant of coke formation was assumed to be constant because of the lack of detailed information as well as to keep the kinetics simple.

An exponential decay function was used to represent the activity drop caused by coking, as shown in Equation 11.8.

$$\ln(k) = \ln(k_0) - \alpha C_c \tag{11.8}$$

where α is the deactivation parameter and C_c is the coke concentration. Hydrogenation and dehydrogenation and cracking reaction rates were the most affected by coking.

11.6.4 Gas Oil Composition

The feed molecular composition was developed using the techniques outlined in Chapter 2. The analytical characterization was transformed into a stochastic feed composition described in terms of structural attributes. All molecules with a mole fraction smaller than 0.005 in the "stochastic feed" were ignored because they are not expected to affect the product distribution significantly. Thus, 222 molecules, including paraffins, naphthenes, aromatics, and hydroaromatics were used to represent the feed.

11.6.5 Model Diagnostics and Results

The various reaction matrices corresponding to each reaction family (Figure 11.4) were applied to all the molecules in the feed and the product to generate a pathways level reaction network. During the model development stage, various candidate models were developed using different reaction rules. The reaction rules listed above were retained in the final model. The reaction network was mathematically represented in terms of ODEs. The system of ODEs was then solved with initial molecular composition and appropriate process conditions, and the rate constants were optimized to experimental data.

Three models were developed to test the pathways level model building capability and the validity of the rules. The models had the same reaction families and reaction rules, but different feeds. The feeds for the models were n-heptane, n-hexadecane, and gas oil. The n-heptane reaction network is a subnetwork of the n-hexadecane reaction network, which is a subnetwork of the gas oil reaction network. The rate constants for the three models were optimized independently, as the available experimental data were for different catalysts. The diagnostics for the three models are listed in Table 11.5. The results are shown in Figure 11.5 to Figure 11.7.

As expected, the solution time for the pathways level models was short. The largest gas oil model was solved in less than 15 CPU seconds on a 400-MHz

TABLE 11.5
Model Diagnostics

Feed	Heptane	Hexadecane	Gas Oil
Number of molecules	15	120	421
Number of reactions	26	526	2021
Paraffin cracking reactions	8	179	596
Paraffin isomerization reactions	3	50	50
Olefin cracking reactions	1	69	87
Olefin isomerization reactions	0	29	29
Hydrogenation reactions	6	62	138
Dehydrogenation reactions	6	59	363
Naphthenic cracking reactions	NA	NA	274
Aromatic cracking reactions	NA	NA	221
Model generation time	7s	38s	427s
Model solution time	<0.1s	≈1s	≈60 s

Pentium II machine. The model results agreed well with the experimental data. The same set of rate constants was used for a wide range of WHSV (Weight Hourly space Velocity) (6.5 to 19.5 for heptane and 71 to 357 for hexadecane). In all cases, the model results were within 10% of the experimental data.

The good agreement between the experimental and model results for a practical range of operating conditions and feedstocks establishes the use of pathways level models as fast and efficient alternatives to mechanistic models.

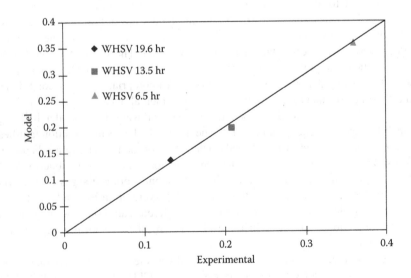

FIGURE 11.5a Parity plot for conversion of heptane FCC model.

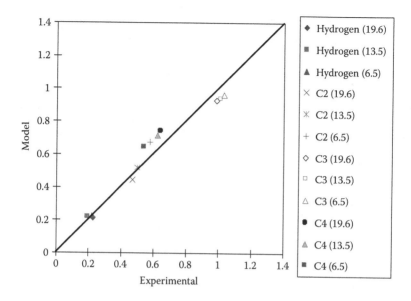

FIGURE 11.5b Parity plot for carbon number distribution for heptane FCC model (numbers in brackets indicate the WHSV).

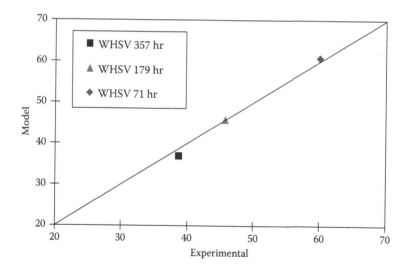

FIGURE 11.6a Parity plot for conversion of hexadecane FCC model.

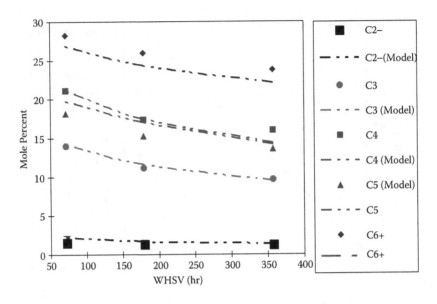

FIGURE 11.6b Parity plot for carbon number distribution of hexadecane FCC model.

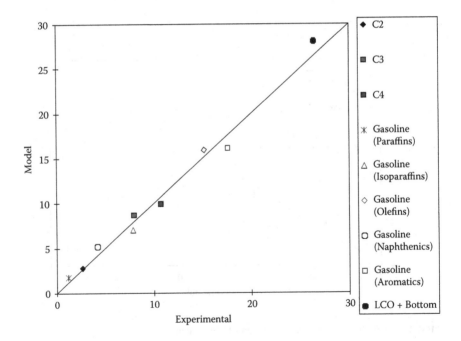

FIGURE 11.7 Parity plot for pathways gas oil FCC model.

11.7 SUMMARY AND CONCLUSIONS

The use of the Kinetic Modeler's Toolbox (KMT) tools for the development of mechanistic and pathways level models for FCC has been demonstrated. The completely automated approach to model building has been demonstrated for pure-component and complex-mixture models.

The concept of stochastic rules has been developed and shown to work effectively in reducing the size of large mechanistic models without significantly affecting the product distribution. However, due to inherent stiffness, the solution of mechanistic models is a very CPU-intensive process even with the use of stochastic rules. As a solution to this problem, the viability of using a faster-solving pathways level model for modeling FCC at the molecular level has been demonstrated.

The automated model building capability for pathways level gas oil FCC models has been developed and applied successfully to both pure components and complex gas oil mixtures. The development of the pathways level model was guided by mechanistic insights to the FCC chemistry.

REFERENCES

Campbell, D., Stochastic Modeling of Structure and Reaction in Hydrocarbon Conversion, Ph.D. dissertation, University of Delaware, Newark, 1998.

Dumesic, J.A., Rudd, D.F., Aparicio, L.M., Rekoske, J.E., and Trevino, A.A., *The Microkinetics of Heterogeneous Catalysis,* American Chemical Society, Washington, DC, 1993.

Dunn, I.J., Linear free energy relationships in modeling heterogeneous catalytic reactions, *J. Catal.,* 12, 335–340, 1968.

Froment, G.E. and Bischoff, K.B., Kinetic data and product distributions from fixed bed catalytic reactors subject to catalyst fouling, *Chem. Eng. Sci.,* 17, 105–114, 1962.

Jacob, S.M., Gross, B., Volts, S.E., and Weekman, V.W., A lumping and reaction scheme for catalytic cracking, *AIChE J.,* 22, 701–713, 1976.

John, T.M. and Wojciechowski, B.W., On identifying the primary and secondary products of the catalytic cracking of neutral distillates, *J. Catal.,* 37, 348, 1975.

Joshi, P.V., Molecular and Mechanistic Modeling of Complex Process Chemistries: A Generic Approach to Automated Model Building, Ph.D. dissertation, University of Delaware, Newark, 1998.

Joshi, P.V., Iyer, S.D., and Klein, M.T., Automated Kinetic modeling of gas oil catalytic cracking, *Rev. Process Chem. Engin.,* 1, 111–140, 1998.

Korre, S.C., Quantitative Structure/Reactivity Correlations as a Reaction Engineering Tool: Applications to Hydrocracking of Polynuclear Aromatics, Ph.D. thesis, University of Delaware, 1994.

Liguras, D.K. and Allen, D.T., Structural models for catalytic cracking. 1. Model compounds reactions, *Ind. Eng. Chem. Res.,* 28(6), 665–673, 1989a.

Liguras, D.K. and Allen, D.T., Structure models for catalytic cracking. 2. Reactions of simulated oil mixtures, *Ind. Eng. Chem. Res.,* 28(6), 674–683, 1989b.

Mochida, I. and Yoneda, Y., Linear free energy relationships in heterogeneous catalysis. I. Dealkylation of alkylbenzenes on cracking catalysts, *J. Catal.*, 7, 386–392, 1967a.

Mochida, I. and Yoneda, Y., Linear free energy relationships in heterogeneous catalysis. II. Dealkylation and isomerization reactions on various solid acid catalysts, *J. Catal.*, 7, 393–396, 1967b.

Mochida, I. and Yoneda, Y., Linear free energy relationships in heterogeneous catalysis. III. Temperature effects in dealkylation of alkylbenzenes on the cracking catalysts, *J. Catal.*, 7, 223–230, 1967c.

Neurock, M. and Klein, M.T., When you can't measure—model, *CHEMTECH*, 23(9), 26–32, 1993.

Quann, R.J. and Jaffe, S.B., Structure oriented lumping: describing the chemistry of complex hydrocarbon mixtures, *Ind. Eng. Chem. Res.*, 31(11), 2483–2497, 1992.

Quann, R.J. and Jaffe, S.B., Building useful models of complex reaction systems in petroleum refining, *Chem. Eng. Sci.*, 51(10), 1615, 1996.

Watson, B.A., Klein, M.T., and Harding, R.H., Mechanistic modeling of n-heptane cracking on HZSM-5, *Ind. Eng. Chem. Res.*, 35(5), 1506–1516, 1996.

Watson, B.A., Klein, M.T., and Harding, R.H., Catalytic cracking of alkylcyclohexanes: modeling the reaction pathways and mechanisms, *Int. J. Chem. Kinet.*, 29(7), 545, 1997.

Weekman, V.W. and Nace, D.M., Kinetics of catalytic cracking selectivity in fixed, moving, and fluid bed reactors, *AIChE J.*, 16, 397–404, 1970.

Yanik, S.J., Demmel, E.J., Humphries, A.P., and Campagna, R.J., FCC catalysts containing shape-selective zeolites to boost gasoline octane number and yield, *Oil Gas J.*, May 13, 108–116, 1985.

12 Automated Kinetic Modeling of Naphtha Pyrolysis

12.1 INTRODUCTION

Hydrocarbon pyrolysis is a major source of ethylene, propylene, butadiene, and aromatics, the principal feedstocks of the petrochemical industry. Thermal cracking of naphtha is becoming increasingly important because of the large yield of olefins it creates. Uncertainties in the petrochemical feedstock market require today's more than billion-pound-per-year pyrolysis plants be designed for a variety of feedstocks. For design of such versatile crackers, an overall kinetic equation for the feed disappearance is generally not sufficient, since the prediction of the product distribution is a very important design aspect. There is now a general trend toward the development and use of detailed kinetic models for a wide range of feedstocks.

Increasing environmental regulations require a detailed characterization of the effluent streams at the molecular level. Questions of unprecedented molecular detail are posed of reaction models. The level of detail sought is incompatible with the contrived and constrained nature of the lumped models. This has provided additional motivation to develop detailed kinetic models.

Developing detailed kinetic thermal cracking models at the mechanistic level is an attractive proposition. Thermal cracking reactions proceed through free radical mechanisms (Laidler, 1965). Many specific mechanisms have been proposed to explain the cracking of simple molecules (Sundaram and Froment, 1978; Dente and Ranzi, 1983). Mechanistic models use elementary reaction steps and their associated kinetics to describe the behavior of the reacting feedstock quantitatively. The main advantages of well-balanced and adequately extensive mechanistic models are as follows:

1. They have the capacity for wide extrapolation and considerable flexibility (in terms of components, mixtures, and operating conditions). After characterization of the types of reactions and their intrinsic kinetic parameters, *a priori* modeling of reactions, rate constants, and product distributions can be developed for previously untested hydrocarbons. For example, starting from a well-tested model of light paraffins

and naphthas, the SPYRO (Dente et al., 1979) model was quickly extended to gas oils with low costs (Goosens et al., 1978a,b)

2. This type of model can accommodate all types of experimental data (bench scale, pilot scale, and commercial).

Developing mechanistic models is now possible because of two enabling advances. First, recent developments in analytical chemistry now permit the direct or at least indirect measurement of the molecular structure of complex mixtures such as naphthas and gas oils, as discussed in Chapter 2. Second, the explosion in computational power allows for the necessary bookkeeping to track the fate of all the molecules during reaction and separation processes, as discussed in Chapter 3. Collectively, the combination of the strategic forces on kinetic models and the enabling analytical and computational advances helped motivate the development of a mechanistic kinetic model for pyrolysis of complex hydrocarbon mixtures such as naphthas and gas oils.

12.2 CURRENT APPROACH TO MODEL BUILDING

Achieving detailed mechanistic model development is not without difficulties. The increase in molecular detail is accompanied by a tremendous increase in both the number of reactions in the governing network and the number of associated rate constants. Even the simplest cases (e.g., pure ethane or propane pyrolysis) require an extensive kinetic scheme in order to represent the real system behavior.

Clearly, some "lumping" scheme that does not sacrifice the mechanistic chemistry is needed. With this in mind, it is useful to realize that the large demand for rate constants in the mechanistic model can be handled by lumping the reactions that involve similar transition states into one reaction family, the kinetics of which is constrained to follow a structure reactivity relationship.

Applying these ideas to pyrolysis starts with an examination of the feedstock. The feedstock can be grouped into a few compound classes such as paraffins, isoparaffins, olefins, naphthenes, and aromatics, which in turn react through a limited number of mechanistic reaction families such as initiation, hydrogen abstraction, β-scission, ring opening and ring closure, radical addition, Diels–Alder condensation, and terminations. As a result, a small number of formal reaction operations can be used to generate many reactions. Further, within a reaction family, differences in the reactivity can be traced to the different types and number of substituents.

Quantitatively, the substituent effects on the reactions within a reaction family can be handled through the use of linear free energy relationships (LFERs). An LFER is a semi-empirical correlation of kinetic parameters with a structural property of the reacting species. The early work of Hammett (1937), Bronsted (Lowry and Richardson, 1987), and Evans and Polanyi (1938) provide the classical formalisms. In the present application to pyrolysis, each reaction family is defined by a single Arrhenius A factor, and its LFER relates the change in the activation energy to a reactivity index. The reactivity index could be a property of one of the molecules or intermediates involved in the reaction, such as heat of formation or carbon number, or a property of the reaction itself, such as the enthalpy change of the reaction. Equation 12.1 summarizes the essential idea:

$$\log k_i = a + \sum_1^n b_i RI_j \qquad (12.1)$$

where i represents a component in a reaction family, j represents a reaction family, RI represents the reactivity index, and the parameters a and b are determined by optimization of experimental data. In the context of pyrolysis chemistry, we use the enthalpy of the reaction as the reactivity index and thus constrain the reactions to follow the Evan–Polanyi relationships.

Large mechanistic models are tedious to build as well as to solve due to the extremely large number of species and their reactions. The computer-assisted automated reaction model generation discussed in Chapter 3 has been employed to generate the current naphtha pyrolysis model. The heat of reaction required for evaluation of the rate constants is done on-the-fly during the reaction network generation. To this end, the Benson's group additivity method as extended by the NIST database (Lias et al., 1994) has been employed to estimate the heat of formation for all species.

12.3 PYROLYSIS MODEL DEVELOPMENT

Pyrolysis elementary reactions are driven by the thermochemistry of radicals and product molecules. Initiation occurs via bond fission of C—C bonds. These give rise to free radicals.

The radicals can then undergo several different reactions. The cracking reaction occurs via β-scission to form an olefinic product and a smaller radical. Hydrogen abstraction produces a stable molecule and a new radical. These steps together represent the propagation steps in a Rice–Herzfeld (Benson, 1980) chain. Termination occurs through radical–radical recombinations and disproportionations that result in the formation of stable species. Most reactions involve radicals, but some purely molecular reactions also play a significant role. Benson (1970) observed that the exclusion of molecular reactions that occur simultaneously with radical reactions has been responsible for some misleading conclusions of rate parameters, particularly for olefins and dienes. The most significant reaction in this category is the Diels–Alder condensation in which an olefin reacts with a conjugated diene to form a cyclic olefin.

The reaction model was generated through the application of the foregoing elementary steps: (a) bond fission, (b) hydrogen abstraction, (c) β-scission, (d) radical addition, (e) Diels–Alder condensation, and (f) termination to every reactant and every species generated during the reaction.

12.3.1 REACTION RULES

Formally, the construction of the model amounted to application of the reaction family matrices to the reactants and subsequent product species. Before applying a particular reaction matrix to a species, the species was tested to check whether it could undergo that reaction. For example, before applying the H-abstraction reaction matrix to a reactant, care was taken to ensure that there existed an abstractable hydrogen, in other words, a check was made to identify all the C—H bonds in the molecule. This was a necessary condition for the reactant to undergo a particular reaction. Additionally, several chemistry-based reaction rules were invoked to guide the model building process. This was to prevent the formation of irrelevant and highly unstable products and keep the number of species and reactions to a minimum without sacrificing chemical detail. For example, in H-abstraction of higher n-alkanes, only C—H bonds belonging to secondary carbon atoms were retained. Thus, a species had to satisfy both the test and rule in order to undergo a particular reaction. In other words, a test is a rigid feasibility requirement that determines whether a reaction is possible for a species. In contrast, a rule allows user knowledge, hypothesis, or biases to filter realized reactions further. Tests are always valid under all conditions of temperature, pressure, and other operating conditions, whereas rules can differ quite substantially from one another depending on the operating conditions.

Various tests and rules that govern the pyrolysis reaction families are described below:

12.3.1.1 Initiation

This is a step in which molecules undergo fission to form radicals.

Test: The molecule must have C—C site.
Rules:
    a. For paraffins less than carbon number 5, initiation is permitted at all C—C bonds.
    b. For n-alkanes with carbon number greater than 5, initiation is permitted only at the central C—C bond.
    c. For branched paraffins, initiation is carried out at the branch point. This results in the formation of a secondary or tertiary radical.
    d. In ringed compounds only side chain cleavage for alkyl naphthenes is allowed.
    e. Formation of bi-radicals through initiation is not permitted.

12.3.1.2 Hydrogen Abstraction

This is a step in which a radical abstracts a hydrogen atom from a molecule to form a molecule and another radical.

Test:

 a. The molecule being attacked by a radical must have at least one C—H bond.

 b. All radicals can undergo this reaction.

Rules:

 a. For linear paraffins with carbon number greater than 6, formation of only secondary radicals is permitted. For lower paraffins, all possible hydrogen abstractions are permitted.

 b. For branched paraffins, hydrogen abstraction is permitted at the branch point. This allows for the formation of the highly stable tertiary radical. We also permit another hydrogen abstraction at a position, which is β to the tertiary position (only if the resultant radical is secondary). The latter can undergo a β-scission reaction to form the stable tertiary radical.

 c. For olefins, hydrogen abstraction is permitted at two positions. One of these results in the formation of the highly stable allylic radical and another results in the formation of a radical which is β to the allylic position.

 d. For ethylene and propylene hydrogen, abstraction is allowed at all possible C—H positions. This is allowed to explain the formation of certain important products such as acetylene, methylacetylene, and allene.

 e. For cyclopentane and cyclohexane, hydrogen abstraction is allowed from all positions. These result in the formation of equivalent radicals. For branched naphthenes, hydrogen abstraction is allowed at the branch point and β to the branch point. For cyclic olefins, hydrogen abstraction is allowed at the allyl position to the double bond. These rules explain dealkylation reactions as well as formation of aromatics from naphthenes.

 f. For aromatic compounds, hydrogen abstraction is allowed to form the highly stable benzylic-type radical. Another hydrogen abstraction that is β to the benzylic position is also allowed.

 g. Dienes were not allowed to undergo H-abstraction reactions.

 d. Only small radicals (carbon number less than 5) are allowed to perform H-abstraction reactions.

12.3.1.3 β-Scission

This is a step in which a radical undergoes unimolecular fission at a bond β to the radical center to form a smaller radical and a stable molecule with a double bond.

 Test: All radicals with a bond β to the radical center can undergo this reaction. Algorithmically, D•-E-F must be present, where D and E must be a carbon atom, whereas F can be a carbon atom or a hydrogen atom.

Rules:

    a. If atom F is a hydrogen atom, then the following rules operate:

      1. If there is a double bond adjacent to the β bond, then β-scission is not permitted.

      2. For cyclic compounds, if atoms D or E (but not both) belong to a cycle, then β-scission is not permitted. This rule prevents the formation of compounds with a double bond adjacent to the ring, as these compounds are not observed in the product spectrum.

      3. For noncyclic radicals, this β-scission is not permitted if the carbon number is greater than 5.

      4. Formation of trienes from dienes is not permitted through this β-scission because no measurable amounts of trienes were seen in the product distribution.

    b. If atom F is a carbon atom, then the following rules apply:

      1. If there is a double bond adjacent to the β bond, then β-scission is not permitted.

      2. For carbon numbers greater than 12, formation of a methyl radical as one of the products of β-scission is not permitted.

      3. For cyclic olefins ring opening is not permitted.

      4b. For multiring naphthenes, ring opening takes place so as to form a single branch.

12.3.1.4 Radical Addition to Olefins

This is a propagation step in which a radical adds to an olefin to form another radical.

Test:

    a. All radicals undergo this type of reaction.

    b. All olefins also undergo this type of reaction.

Rules:

    a. Since it is a bimolecular reaction, the rate of reaction drops drastically with increase in radical size (Benson, 1980). Hence, molecules and radicals with carbon number greater than 5 do not undergo this reaction.

    b. The carbon number of the products must not exceed six.

12.3.1.5 Diels–Alder Reaction

This is a molecular reaction that takes place between a diene and an olefin to form a cyclic alkene.

Test:
 a. All conjugated dienes can undergo this reaction.
 b. All olefins can undergo this reaction.
Rules:
 a. Only dienes, butadienes, and pentadienes undergo this reaction.
 b. Only ethylene and propylene undergo this reaction.

For light paraffin cracking, Diels–Alder reactions play an important role in explaining the formation of aromatics. However, for the heavy feedstocks such as naphtha and gas oils, they do not have a significant effect on the product distribution since the extent of reaction is fairly low. The Diels–Alder reactions have been included in the final model for completeness.

12.3.1.6 Termination Reactions

This is a radical reaction in which two radicals react with each other to form a stable molecule.

Test: All radicals can undergo this type of reaction.
Rules: Only small radicals are allowed to undergo this type of reaction (carbon number less than 3). Since this is a bimolecular reaction, bigger radicals (which are generally slower) have less likelihood of undergoing this reaction.

The reaction type, reaction matrices, and rules for each reaction family are summarized in Table 12.1.

12.4 CONTRIBUTION OF REACTION FAMILIES

It is instructive and useful to consider the role of the various reaction families in the thermal cracking chemistry. The β-scissions and hydrogen abstractions are the fastest reaction families, and most of the product species are a direct result of these reactions. The radical addition to olefin reactions has relatively minor importance in terms of the extent of reaction. It is nonetheless critical for correctly predicting the product distribution.

The radical addition to olefins provides the dominant pathway for the generation of branched species from normal species and larger species from smaller ones. For example, in the case of n-octane cracking, the product has nonzero concentrations of isoheptane, iso-octane, and undecanes. These larger branched isomers through successive β-scission crackings produce significant amounts of smaller molecules including methane and isobutane. Thus, though radical addition reactions have smaller rate constants, they can have a significant effect on the product distribution.

TABLE 12.1
Reaction Matrices and Rules for Mechanistic Reaction Model of Naphtha Pyrolysis

Reaction Family	Reaction Matrix	Reaction Rules
Bond fission $R-C_1H_2-C_2H_2-R' \longrightarrow R-\dot{C_1}H_2 + R'-\dot{C_2}H_2$	$\begin{array}{c c} & \begin{array}{cc} C_1 & C_2 \end{array} \\ \begin{array}{c} C_1 \\ C_2 \end{array} & \begin{bmatrix} 1 & -1 \\ -1 & 1 \end{bmatrix} \end{array}$	1. Allow only central bond fission for large n-paraffins 2. Allow fission at branch point for branched paraffins
Hydrogen abstraction by radical $R-H + \dot{R}' \longrightarrow R'-H + \dot{R}$	$\begin{array}{c c} & \begin{array}{ccc} R & H & \dot{R}' \end{array} \\ \begin{array}{c} R \\ H \\ \dot{R}' \end{array} & \begin{bmatrix} 0 & -1 & 0 \\ 1 & 0 & 1 \\ 0 & 1 & -1 \end{bmatrix} \end{array}$	1. Allow formation of only secondary radicals for higher n-alkanes 2. Allow H-abstraction at branch points and β to branch for isoparaffins 3. Allow H-abstraction at allylic position and β to allylic position for olefins 4. For branched napthenes allow H-abstraction at branch and β to branch 5. For ethylene and propylene allow all H-abstractions 6. For aromatic compounds allow formation of stable benzylic species 7. Allow small radicals to attack (carbon number <5)

TABLE 12.1(Continued)
Reaction Matrices and Rules for Mechanistic Reaction Model of Naphtha Pyrolysis

Reaction Family	Reaction Matrix	Reaction Rules

β-Scission

$R'{-}CH_2{-}\overset{\bullet}{R} \longrightarrow R{=}CH_2 + \overset{\bullet}{R'}$

$$\begin{array}{c} \\ R \\ C \\ {}^{\bullet}R' \end{array} \begin{array}{ccc} R & C & \overset{\bullet}{R'} \\ \begin{bmatrix} -1 & 1 & 0 \\ 1 & 0 & -1 \\ 0 & -1 & 1 \end{bmatrix} \end{array}$$

1. Do not allow formation of double bond adjacent to ring
2. Restrict formation of trienes from dienes
3. Ring opening not permitted for cyclic olefins
4. Formation of H• not permitted for carbon number >5; formation of CH_3 not permitted for C >12
5. For multiring compounds, ring opening results in a the formation of a single branch

Radical Addition to Olefins

$\overset{\bullet}{R} + R'{-}C_1H{=}C_2H_2 \longrightarrow R'{-}\overset{\bullet}{C_1}H{-}C_2H_2{-}R$

$$\begin{array}{c} \\ \overset{\bullet}{R} \\ C_1 \\ C_2 \end{array} \begin{array}{ccc} \overset{\bullet}{R} & C_1 & C_2 \\ \begin{bmatrix} -1 & 1 & 0 \\ 1 & 0 & -1 \\ 0 & -1 & 1 \end{bmatrix} \end{array}$$

1. Allow small radicals to attack (carbon number <5)
2. Restrict the product size (carbon number <6)

Diels–Alder Condensation

$$\begin{array}{c} \\ C_1 \\ C_2 \\ C_3 \\ C_4 \\ C_5 \\ C_6 \end{array} \begin{array}{cccccc} C_1 & C_2 & C_3 & C_4 & C_5 & C_6 \\ \begin{bmatrix} 0 & -1 & 0 & 0 & 1 & 0 \\ -1 & 0 & 1 & 0 & 0 & 0 \\ 0 & 1 & 0 & -1 & 0 & 0 \\ 0 & 0 & -1 & 0 & 0 & 1 \\ 1 & 0 & 0 & 0 & 0 & -1 \\ 0 & 0 & 0 & 1 & -1 & 0 \end{bmatrix} \end{array}$$

1. Only conjugated butadienes and pentadienes can undergo this reaction
2. Only C2 and C3 olefins can undergo this reaction

(*Continued*)

TABLE 12.1(Continued)
Reaction Matrices and Rules for Mechanistic Reaction Model
of Naphtha Pyrolysis

Reaction Family	Reaction Matrix	Reaction Rules
Radical Termination $\dot{R}$ + $\dot{R}'$ $\longrightarrow$ R —— R′	$\begin{array}{c} \\ \dot{R}- \\ \dot{R}' \end{array}\begin{array}{cc} \dot{R} & \dot{R}' \\ \left[\begin{array}{cc} -1 & 1 \\ 1 & -1 \end{array}\right] \end{array}$	1. Allow terminations for carbon number <3 2. All termination products must have pathways to react further

The initiation and the termination reactions have a more complex role in thermal cracking chemistry. For most cases of practical interest, initiation and termination reactions have a small direct contribution to the product distribution but are important in terms of generating and maintaining the radical pool for sustaining the important β-scissions and hydrogen abstractions.

The radical concentrations in most industrial applications are smaller by a few orders of magnitude than the concentration of the molecules. The radicals are unstable intermediates and have large rate constants for cracking (β-scissions) into olefins and smaller radicals. For large molecules, the β-scissions are perhaps the fastest reactions in thermal cracking chemistry. The competing pathways of H-abstraction, olefin addition, and radical recombination have a significantly smaller rate of reaction for larger radicals that can undergo one or more β-scissions. The molar concentration of larger radicals (C5+, which have viable β-scissions pathways) is much lower compared to, for example, ethyl or propyl radicals (which cannot undergo further β-scissions). For most cases of practical interest, more than 90% of the radical pool is composed of radicals with four or fewer carbons. This provides a semitheoretical justification for allowing only radicals with fewer than five carbons to participate in hydrogen abstraction and olefin addition reactions. Empirical evidence suggests that this does not lead to any significant loss of the chemical significance of the model.

12.5 REACTION NETWORK DIAGNOSTICS

Application of the reaction matrices in Table 12.1, with the associated rules, generated a mechanistic model totaling 7774 reactions that could be organized into reaction families as follows: 174 initiation reactions, 6722 hydrogen abstraction reactions, 675 β-scission reactions, 92 addition reactions including Diels–Alder condensation reactions, and 21 termination reactions. Ring closure and ring opening reactions were considered implicitly as a subset of β-scission reactions. It should be noted that reaction path degeneracy was accounted for in the kinetics of the model.

It also formed 799 different species as products of the reaction (including 11 n-alkanes; 34 isoparaffins; 72 olefins including ethylene and propylene; 38 dienes including allene, 1-3-butadiene, and pentadiene; 106 naphthenes including cyclic olefins and cyclic diolefins; 26 aromatics including benzene, toluene, xylenes, and styrene; hydrogen; acetylene; methylacetylene; and 512 radicals). The change in concentration for each species was represented by an ordinary differential equation. The full set of 799 ordinary differential equations constituted the naphtha mechanistic model for pyrolysis. The solution of this set of 799 equations required estimates of the rate constants for each elementary step, the source being the LFER.

12.6 PARAMETER ESTIMATION

In Section 12.2, the rate constants within each reaction family were described in terms of a family-specific Arrhenius A factor and the Polanyi relationship parameters that relate the activation energy to the enthalpy change of reaction, as shown in Equation 12.2:

$$E_A = E_o + \alpha * \Delta H_{rxn} \qquad (12.2)$$

Equation 12.3 shows how the Polanyi relationship and the Arrhenius expression were combined to represent the rate constant, k_{ij}, where i denotes the reaction family and j denotes the specific reaction in the family:

$$k_{ij} = A_i + \exp(-(E_{o,i} + \alpha_i * \Delta H_{rxn,j}) / RT) \qquad (12.3)$$

In short, each reaction family can be described with a maximum of three parameters (A, Eo, α). Procurement of a rate constant from these parameters requires only an estimate of the enthalpy change of reaction for each elementary step. In principle, this enthalpy change of reaction amounts to the simple calculation of the difference between the heats of formation of the products and reactants. However, since many model species, particularly the radical intermediates, are without experimental values, an estimation method is required. For the work described here, group additivity methods were used (Lias et al., 1994) to evaluate the heats of formation of every species. For consistency, group additivity–calculated values of ΔH_f were utilized even when experimental values were available.

Therefore, in principle, the thermal cracking model described thus far would have three LFER parameters per reaction family, A, E_o, and α. However, the total number of adjustable parameters to be determined through fitting the model to the data was much smaller. For example, most A factor values were taken from literature estimates that used transition theory arguments

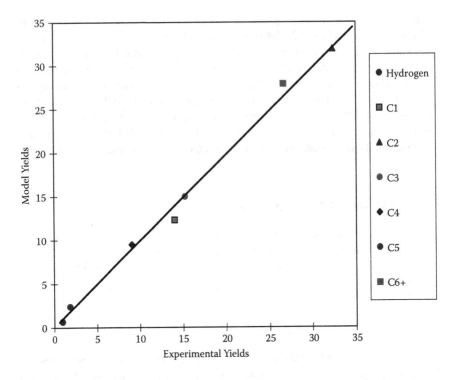

FIGURE 12.1 Parity plot for product distribution by carbon number of naphtha pyrolysis.

(Benson, 1980). Also, the value of α was fixed at the literature values for most reaction families.

Pilot plant data were used to optimize the 799 component models by using the GREG local optimization program (Stewart et al., 1992) to tune the LFER parameters together with the structural parameters The predicted yields were then compared to the experimental data through a typical χ^2 objective function.

The results of the model regression of the naphtha data are summarized in Figure 12.1 through Figure 12.3 in the form of parity plots. The model predictions agree reasonably well with the experimental data.

12.7 SUMMARY AND CONCLUSIONS

A mechanistic model used to describe the thermal cracking of complex hydrocarbon mixtures (e.g., naphtha) has been developed. The model was based on the concept of reaction families and associated LFERs for rate equation evaluation.

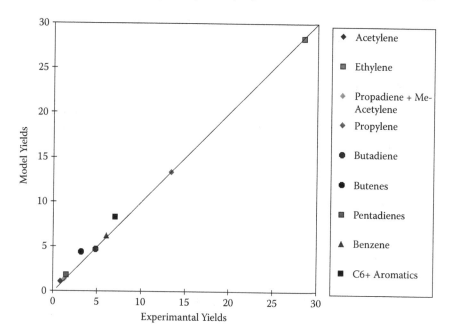

FIGURE 12.2 Parity plot for unsaturated hydrocarbon products of naphtha pyrolysis.

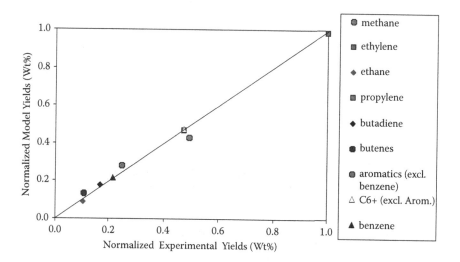

FIGURE 12.3 Parity plot for low yield products of naphtha pyrolysis (all yields are normalized with the yield of ethylene).

This lumping scheme was used to obtain molecular level detail without requiring extensive experimental data and rate constant values.

REFERENCES

Arai, Y., Murata, M., Tanaka, S., and Saito, S., Simulation of product distributions from pyrolysis of normal and branched alkane mixtures over a wide range of conversions, *J. Chem. Eng. Jpn.,* 10, 303–307, 1977.

Benson, S.W., Refining petroleum for chemicals, In *Advances in Chemistry Series,* No. 97, Chapter 1, American Chemical Society, Washington, DC, 1970.

Benson, S.W., The thermochemistry and kinetics of gas phase reactions, In *Frontiers of Free Radical Chemistry,* Academic Press, New York, 1980.

Broadbelt, L.J., Stark, S.M., Klein, M.T., Computer generated pyrolysis modeling: on-the-fly generation of species, reactions and rates, *Ind. Eng. Res. Chem.,* 33, 790–799, 1994.

Campbell, D.M. and Klein, M.T., Construction of a molecular representation of a complex feedstock by Monte-Carlo and quadrature methods, *Appl. Catal.,* AGEN 160, 41–54, 1997.

Davis, H.G. and Williamson, K.D., In *Thermal Hydrocarbon Chemistry,* Product Inhibition in the Pyrolysis of Paraffinic Hydrocarbons Oblad, A.G., Davis, H.G., and Eddinger, R.T., Eds., Advances in Chemistry Series, No. 183, Chapter 4, Am. Chem. Soc., 41-66, Washington, DC, 1979.

Dente, M.E. and Ranzi, E.M., Mathematical modeling of pyrolysis reactions, In *Pyrolysis: Theory and Industrial Practice,* Albright, L.F., Crynes, B.L., and Corcoran, W.H., Eds., Academic Press, New York, 1983.

Dente, M., Ranzi, E., Goossens, and A.G., Detailed prediction of olefin yields from hydrocarbon pyrolysis through a fundamental simulation model (SPYRO), *Comput. Chem. Eng.,* 3, 61–75, 1979.

Evans, M.G. and Polanyi, M., Inertia and driving force of chemical reactions. *Trans. Faraday. Soc.,* 34, 11, 1938.

Goosens, A.G., Dente, M., and Ranzi, E., Improved steam cracker operation, *Hydrocarbon Process.,* 57(9), 227–236, 1978a.

Goosens, A.G., Dente, M., and Ranzi, E., Simulation program predicts olefin-furnace performances, *Oil Gas J.,* 76(36), 89–104, 1978b.

Hammett, L.P., The effect of structure upon the reactions of organic compounds, *J. Am. Chem. Soc.,* 59, 96–103, 1937.

Laidler, K.J., *Chemical Kinetics.,* 2nd ed., McGraw-Hill, New York, 1965.

Lias, S.G., Liebman, J.F., Levin, R.D., and Kafafi, S.A., *NIST Standard Reference Database 25, Structure and Properties,* gaithersbury, MD, 1994.

Lowry, T.H. and Richardson, K.S., *Mechanism and Theory in Organic Chemistry,* 3rd ed., Harper and Rowe, New York, 1987.

Murata, M. and Saito, S., Simulation model for high-conversion pyrolysis of normal paraffinic hydrocarbons, *J. Chem. Eng. Jpn.,* 8, 39–45, 1975.

Stewart, W.E., Caracotsios, M., and Sorenson, J.P., Parameter estimation from multiresponse data, *AIChE J.,* 38(5), 641–650, 1992.

Sundaram, K.M. and Froment, G.F., Modeling of thermal cracking kinetics, *Chem. Eng. Sci.,* 32, 601–608, 1977a.

Sundaram, K.M. and Froment, G.F., Modeling of thermal cracking kinetics-II, *Chem. Eng. Sci.*, 609–617, 1977b.

Sundaram, K.M. and Froment, G.F., Modeling of thermal kinetics. 3. Radical mechanisms for the pyrolysis of simple paraffins, olefins and their mixtures, *Ind. Eng. Chem. Fundam.*, 17, 174–182, 1978.

Van Damme, P.S., Narayanan, S., and Froment, G.F., Thermal cracking of propane and propane-propylene mixtures: pilot plant versus industrial data, *AIChE J.*, 21, 1065–1073, 1975.

13 Summary and Conclusions

13.1 SUMMARY

This book has developed both a general chemical engineering kinetic modeling toolbox (KMT) and a set of applications that demonstrate its utility. The molecule-based kinetic modeling strategy has been developed and automated to build rigorous fundamental models for complex reaction systems. The model building, solving, optimization, and delivery techniques have been exploited systematically. This integrated approach has been successfully applied to modeling various petroleum refining processes, including catalytic reforming, hydrocracking, hydrotreating, hydroprocessing, fluid catalytic cracking (FCC), and pyrolysis. The generic modeling tools can be easily extended to other complex reaction systems and speed up the model development process significantly.

Figure 13.1 provides a three-dimensional view of the coverage of this book. The x-axis summarizes the major theoretical contributions in the systematic molecule-based detailed kinetic modeling approach. The graph-theoretical concepts were used to automate the reaction network building, and the linear free energy relationships (LFERs) or quantitative structure reactivity correlations (QSRCs) were used to organize and estimate rate parameters. The combination of these two formed the kernel of a kinetic model template. To model, end to end, from the complex feedstock characterization to product properties, it uses submodels of structure, reaction, and reactivity. This begins with the stochastic modeling of molecular structure and composition, MolGen, which converts the feedstock to a set of representative molecular structures. The network generator, NetGen, enables us to build a reaction network that can capture the essential chemistry and kinetically significant species. EqnGen then converts the generated network to a system of mathematical equations. The system of equations can be solved efficiently by exploring the inherent stiffness and sparseness properties of the system with SolGen. This modeling process and the generated models can be solved within a parameter optimization framework, ParOpt, to tune the model parameters by matching model results with experimental data.

The y-axis summarizes the work to integrate the above molecule-based kinetic modeling process into a complete system of software tools — the KMT, which has been ported from a legacy server/UNIX platform to a PC/Windows platform. Various model delivery technologies, such as the user interface and documentation, were developed to make the software tools and models easily accessible to end users, process chemists, and engineers.

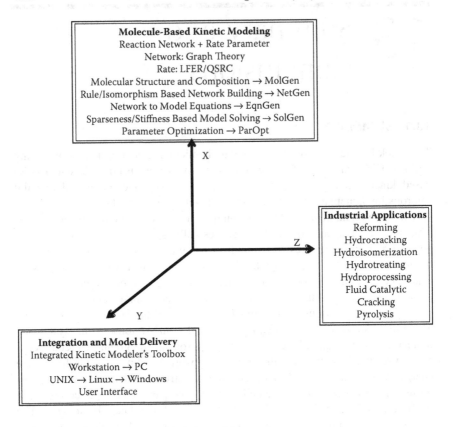

FIGURE 13.1 A three-dimensional view of this work.

The x-axis and y-axis combined form the first part of this book. Their applications are summarized in the z-axis, which forms the second part of this book. The automated molecule-based kinetic modeling approach and KMT were successfully applied to industrial complex processes, including catalytic reforming, hydrocracking, hydrotreating, hydroprocessing, and FCC, as well as pyrolysis. The generic modeling tools have also been extended to other complex reaction systems, such as hydroisomerization, alkylation, and oxidation.

In the following sections, we summarize, chapter by chapter, the important results obtained in this book.

13.1.1 Molecular Structure and Composition Modeling of Complex Feedstocks

1. A quadrature stochastic sampling method to transform the basic analytical chemistry for a complex feedstock into a small set of representative molecules has been developed. The molecules are defined as a

collection of attributes, whose values are represented by a probability distribution function. For a petroleum feedstock, gamma distributions represented the probability of the attribute values well. The information in these probability distribution functions (PDFs) can typically be preserved in a set of $O(10)$ molecules.

2. The more analytical information is available, the better this molecular structure modeling approach works. The objective function used to optimize the PDF parameters and the quadrature mole fractions is flexible to customize, and any analytical information can be accommodated into it to improve the accuracy of the model prediction.

3. It is important to balance time issues with accuracy issues to better develop a statistical representation of a complex feedstock by constraining the representation to a set of analytical characterizations. More detailed distributions to define the structural attributes and larger stochastic representations require more CPU time, so that minimizing CPU time while maintaining the desired level of accuracy should be the primary objective.

4. A solution to obtain a generic molecular representation for each kind of feedstock is to sum all the optimal representative molecules together for a set of typical feeds by applying the modeling approach developed here; then only the mole fractions are needed to further optimize to match the analytical properties.

13.1.2 Automated Reaction Network Building of Complex Process Chemistries

1. The graph-theoretical concepts are utilized to enable automated reaction network construction. A molecule can be represented by a chemical graph, with the atoms being the nodes and the bonds being the edges. The bond–electron matrix representation enables chemical reaction to be implemented through a simple matrix addition operation. Through repetitive application of reaction matrices to the reactants and their progeny, the reaction network can then be built.

2. The network can easily grow beyond the user constraints and even the computer capabilities (CPU, memory, and compiler) when the automated reaction network building algorithms are directly applied to high-carbon-number reactant compounds or complex multicomponent mixtures. The challenge is building the reaction network that can capture the essential chemically and kinetically important species and reactions while keeping the network to a modest size to fulfill our need.

3. Every stage of the model building process is exploited, and a system of methodologies was developed to guide the building of the reaction network. In the preprocessing stage, both the rule-based model building strategy and the seeding and deseeding strategy are exploited and

developed to provide users the flexibility to guide the model building process with their expertise. In the *in situ* processing stage, we extended the original isomorphism algorithm and created the "on-the-fly" generalized isomorphism-based lumping algorithm and the stochastic sampling algorithm. In the postprocessing stage, we used the species-based or reaction-based model reduction algorithms and the isomorphism-based late lumping strategy to reduce the network size.

4. The generalized isomorphism-based lumping strategy fundamentally superseded the classical lumping schemes and provided users the maximum flexibility of building molecule-based reaction models on the basis of fundamental chemistry and reaction mechanism by balancing needs with the available data and QSPRs. This also helps answer the fundamental question of how similar or different the two reaction networks are by comparing two reaction mechanisms from an isomorphic-lumping point of view.

5. The keys to building a useful kinetic model for complex process chemistry are (a) a fundamental chemistry and reaction mechanism, (b) a molecular basis, and (c) being consistent and working. These goals have guided the successful development of automated reaction network construction algorithms and methodologies in this work.

13.1.3 KINETIC RATE ORGANIZATION AND EVALUATION OF COMPLEX PROCESS CHEMISTRIES

1. The rate laws of the complex reaction network are reviewed at both the pathways level and mechanistic level. At the mechanistic level, the rate law is as simple as the mass action rate for each elementary step.

2. The LFER concept for organizing kinetic data is reviewed from the classical transition state theory, and the reaction family is introduced. Representative results developed for catalytic hydrocracking, including adsorption, hydrogenation, isomerization, ring opening, and dealkylation, are presented to demonstrate the wide range of applicability of the LFER concept. A comprehensive set of LFERs developed for catalytic hydrocracking is also organized and tabulated for easy reference.

13.1.4 MODEL SOLVING TECHNIQUES FOR DETAILED KINETIC MODELS

1. The nature of the detailed kinetic models (DKM) and the mathematical background to solve these systems, including the underlying numerical methods, the stiffness of the DKM systems, and the sparseness of the DKM systems, are reviewed.

2. Various solvers are exploited to understand the underlying numerical methods to solve stiff and nonstiff ordinary differential equation

(ODE)/differential algebraic equation (DAE) systems. The sparsity structure of the Jacobian matrix is exploited to improve the performance and accuracy of the numerical solutions. In our experiments, for the pathways level DKM, the solver performance is ordered as Livermore Solvers for Ordinary Differential Equations (LSODES) with Adams > LSODE with Adams > LSODA >> LSODES with backward differentiation formula (BDF) >> LSODE with BDF >> differential/algebraic system solver (DASSL); for the mechanistic level DKM in this experiment, the solver performance is ordered as LSODES with BDF and user-supplied Jacobian > LSODES with BDF and finite-differenced Jacobian >> LSODE with BDF and user-supplied Jacobian > DASSL > LSODE with BDF and finite-differenced Jacobian > LSODA.

3. It is very important to judiciously select appropriate numerical solvers for large DKMs, to exploit the stiffness of the specific physical problem, and to take advantage of the sparsity of large reaction network. All of this can make significant differences in model solving performance. Generally, pathways level DKMs are nonstiff and the Adams–Moulton algorithm is preferred; mechanistic level DKMs are stiff, and for these systems, the BDF algorithm is preferred. DKMs are generally sparse in nature; therefore, a solver such as LSODES that treats the Jacobian matrix's sparsity structure is recommended.

13.1.5 INTEGRATION OF DETAILED KINETIC MODELING TOOLS AND MODEL DELIVERY TECHNOLOGY

1. All the technical components of the detailed kinetic modeling tools are integrated together into one user-friendly software package—the KMT. The KMT has five modules to automate the molecule-based kinetic modeling process, in the following order: the molecule generator (MolGen), the reaction network generator (NetGen), the model equation generator (EqnGen), the model solution generator (SolGen), and the parameter optimization framework (ParOpt). This automated kinetic modeling process enables the modeler to focus on the fundamental chemistry and speed up the model development significantly.

2. The integrated KMT in Figure 6.1 clearly depicts the molecule-based kinetic modeling approach from the input of feedstock characterization to predict the output of product properties. The set of the PDF parameters is just like the fingerprint of a complex feedstock and can be optimized to assemble its molecular representation. The understanding of the reaction chemistry provides the reaction families and reaction rules and the understanding of the reaction kinetics provides us with the rate law. The set of QSRC and LFER parameters

fundamentally correlates the molecular structures with the molecular reactivities.

3. Issues related to KMT development and model delivery are reviewed from the software development perspective and include the platform, porting, data, and documentation issues. The rules for success in developing such a large software system are documented, and a new three-tier Web-enabled distributed architecture for the next generation of KMT software is proposed.

13.1.6 MOLECULE-BASED KINETIC MODELING OF NAPHTHA REFORMING

1. The graph-theoretical approach and the reaction family concept are successfully applied to molecule-based kinetic modeling of the bifunctional heterogeneous catalytic naphtha reforming process. The reaction matrices and reaction rules are reviewed for each reaction family and are tabulated for easy reference. The automated model building capabilities are extended to the naphtha reforming at the pathways level that enables us to build molecular models fast and accurately for a variety of "what-if" scenarios.

2. A C14 naphtha catalytic reforming model with 147 species and 587 reactions was built, optimized, and provided excellent parity between the predicted and experimental yields for a wide range of feedstocks and operating conditions. This demonstrated the feedstock-independent nature of the developed detailed kinetic model.

3. Various chemical insights were drawn by analyzing the subfamilies divided via mechanistic understanding of naphtha reforming. The inclusion of subfamilies in the dehydrocyclization (DHC), isomerization, and hydrocracking reactions made it possible to incorporate mechanistic ideas in the pathways level model. It was found that the stability of the carbenium ions involved in the reaction greatly affects the rate of reaction, which is the key factor to divide the reaction families into subfamilies.

13.1.7 MECHANISTIC KINETIC MODELING OF HEAVY PARAFFIN HYDROCRACKING

1. Graph-theoretic concepts, the reaction family concept, and the QSRC formalism are used to construct a paraffin hydrocracking mechanistic model builder incorporating bifunctional catalysis with metal and acid functions. The paraffin hydrocracking chemistry is described with various reaction families incorporating the metal function (dehydrogenation and hydrogenation) and the acid function (protonation and

deprotonation, H/Me-shift, protonated cyclopropane (PCP) isomerization, and β-scission).

2. A C16 paraffin hydrocracking model with 465 species and 1503 reactions at the mechanistic level was built and optimized. This model provided excellent parity between the predicted and experimental yields for a wide range of operating conditions. This shows the fundamental nature (feedstock and catalyst acidity independent) of the rate parameters in the model.

3. Various insights are obtained from the optimization results of the detailed C16 model: the skeletal isomerizations precede the cracking reactions; PCP isomerization always leads to branching, A-type cracking always leads to branched isomers, and B-type cracking always leads to normal or branched isomers; all the cracking products normally come from A-type cracking of tri-branched isomers or B-type cracking of di-branched isomers.

4. The generalized isomorphism algorithm was applied at the carbon and branch number level to reduce the complexity and explosion of modeling heavy paraffin hydrocracking. A C80 paraffin hydrocracking model with 306 species and 4671 reactions at the pathways level reduced from the fundamental chemistry and reaction mechanism was developed.

13.1.8 MOLECULE-BASED KINETIC MODELING OF NAPHTHA HYDROTREATING

1. The graph-theoretical approach and the reaction family concept are successfully applied to the molecule-based kinetic modeling of the heterogeneous catalytic hydrotreating process. The reaction matrices and reaction rules are reviewed for each reaction family and tabulated for further reference. The automated model building capabilities are exploited and further extended to the hydrotreating at the pathways level and enable us to build rigorous molecular models fast and accurately.

2. It is necessary to incorporate all the representative molecular structures with substituents at both significant and nonsignificant positions in order to rigorously model the HDS chemistry. It is also very important to incorporate dual-site mechanisms and implement the corresponding Langmuir–Hinshelwood–Hougen–Watson (LHHW) formalism to take into account the inhibitions of various compounds in the process stream (especially the H_2S inhibition to the direct desulfurization sites). A structural approximation concept is introduced to account for the steric and electronic effects of substituents on thiophenic compounds to the kinetic reaction rates compared with their nonsubstituted parent molecule.

3. A naphtha hydrotreating model with 243 species and 437 reactions was built automatically and optimized. This model provided excellent parity between the predicted results and pilot plant data for a wide range of operating conditions.

4. Various insights into the naphtha hydrotreating process are described. Olefin hydrogenation has fast first-order kinetics. An aromatic saturation reaction is not significant for the naphtha range hydrotreating since most of the aromatics are one-ring alkylbenzenes, and their saturation rate is slow. Different sulfur compounds demonstrate different reaction rates: mercaptans, sulfides, and disulfides react fast; thiophenes and benzothiophenes are the next to be removed; dibenzothiophenes, especially those with alkyl chains on the 4 and 6 positions are the most difficult to remove. The overall sulfur reaction demonstrates a second-order conversion rate, although each reaction step for each sulfur compound is modeled as a first-order reaction. There exists an optimum temperature for the best selectivity for the sulfur removal versus olefin hydrogenation, and increasing LHSV is also beneficial to this sulfur removal selectivity.

5. This fast-solving naphtha hydrotreating model, which takes only 1 CPU second on an Intel Pentium II 333-MHz machine to run once through, can be used to optimize the low-sulfur selective hydrotreating process quantitatively and be further utilized for process control.

13.1.9 AUTOMATED KINETIC MODELING OF GAS OIL HYDROPROCESSING

1. The automated detailed kinetic modeling capability was significantly enhanced and successfully extended to the gas oil hydroprocessing process with all the process chemistry modules fully integrated, including hydrocracking chemistry for aromatics, hydroaromatics, naphthenes, paraffins, and olefins, HDS chemistry for sulfur-containing compounds, and hydrodenitrogenation (HDN) chemistry for nitrogen-containing compounds. The major enhancements in the model building tools lie in their capability to handle various multiring aromatic, hydroaromatic, naphthenic, and heterocyclic compounds and the new reaction families including ring saturation, ring isomerization, ring opening, ring dealkylation, ring closure, side chain cracking, sulfur saturation, nitrogen saturation, desulfurization, and denitrogenation reactions.

2. A comprehensive molecule-based heavy gas oil hydroprocessing kinetic model with 534 species and 1727 reactions has been developed and can be easily rebuilt and customized with new experimental data. The model showed good agreement with the experimental observations

and the right trends of the molecular conversions in the process stream and can be further tuned and optimized with more available data.

3. This automated kinetic modeling capability for gas oil hydroprocessing has shifted the modeling strategy of this complex process and provides flexibility to allow the users to build various candidate models fast, test out various scenarios, optimize the model adaptively, and select the best model. This development process also demonstrated that the generic kinetic modeling tools can be extended to handle complex reaction systems and significantly speed up the kinetic model development process.

13.1.10 MOLECULAR MODELING OF FLUID CATALYTIC CRACKING

1. The use of graph-theoretic concepts and the completely automated approach to model building for the development of both mechanistic and pathways level models for fluid catalytic cracking have been demonstrated for both pure component and complex mixture models.

2. The concept of stochastic rules has been developed and shown to work effectively in reducing the size of large mechanistic models without significantly affecting the product distribution. However, the solution of mechanistic models, due to their inherent stiffness, is a very CPU intensive process, even with the use of stochastic rules.

3. The automated model building capability for pathways level gas oil FCC model is developed and applied successfully to both pure components and complex gas oil mixtures. The development of the pathways level model is guided by mechanistic insights to the fluid catalytic cracking chemistry.

13.1.11 AUTOMATED KINETIC MODELING OF NAPHTHA PYROLYSIS

A mechanistic level model to describe the thermal cracking of naphtha is developed. The model predictions and experimental data are in excellent agreement, validating all the rules and assumptions that went into the construction of the model.

13.2 CONCLUSIONS

The KMT molecule-based kinetic modeling approach has delineated and reduced the essential elements of the complexity in the complex reaction systems to a manageable and irreducible level: the molecular structures are reduced to O(10)

PDF parameters (five to ten PDFs times two to three parameters in each PDF), the reaction network building is automated through $O(10)$ reaction matrices (normally one for each reaction family), and the molecular reactivities are correlated with $O(10)$ kinetic parameters (five to ten LFERs times two to three parameters each). The brute force description of these complex systems could have been of the order $O(10^5 \times 10)$ due to the $O(10^5)$ species and $O(10)$ reactions for each species in these complex reaction systems, which is prohibitive to construct manually.

All the chemical engineering technical components for detailed kinetic modeling are integrated into one user-friendly software package — the KMT — to automate the modeling of molecular structures and the kinetics of complex reaction systems. The KMT includes five modules: MolGen for molecular structure and composition modeling, NetGen for automated reaction network generation, EqnGen for automated code and equation generation, SolGen for the model solution, and the parameter optimization framework, ParOpt. The KMT software has automated the process of building these detailed kinetic models. By exploiting Monte Carlo and graph-theory techniques, reaction models containing thousands of species can be built in 1000 CPU seconds or less. This incredible model building speed has changed the serial model building–model use paradigm to a new parallel approach, where a model builder can produce an updated optimal model in seconds. The thus-constructed models can then react to the molecularly explicit feedstock using QSRCs for kinetic parameters to predict the product properties.

This integrated approach has been successfully applied to modeling various petroleum refining processes, including catalytic reforming, hydrocracking, hydrotreating, hydroprocessing, and FCC, as well as pyrolysis for feedstocks ranging from pure components to complex mixtures such as naphtha and gas oil. The models match the pilot-plant and commercial data very well and provide insights and quantitative information that can be used for process development and design. The generic modeling tools can be easily extended to other complex reaction systems and speed up the model development process significantly as demonstrated in this book.

Index

Printed in the United States
by Baker & Taylor Publisher Services